Code Brown / Graeme & Robert Parker

Code Brown

written by
Graeme & Robert Parker

First published in Great Britain in 2025
by The Hoof GP
Laigh Kirkland, Wigtown, Scotland
www.thehoofgp.com

Publishing Services from Lumphanan Press
www.lumphananpress.co.uk

Paperback ISBN 978-1-0685215-3-9
Ebook ISBN 978-1-0685215-4-6

Typeset in ATF Garamond
Printed by Short Run Press, Exeter, Devon, UK

*This is a book of humorous anecdotes from the author's life and work. It describes
the author's recollections and honest opinions of events and experiences which
happened over time. Some names have been changed, some events have been
compressed, and some dialogue has been recreated from memory.*

For Ashley – you wanted fifty shades of grey;
I gave you a hundred shades of brown.

Contents

11   Introduction

15   Chapter One: Sound Bites

29   Chapter Two: The Calm Before Wotsisname

45   Chapter Three: Big Boy Boots

63   Chapter Four: Spring is on the March

75   Chapter Five: The April Fools' Club

98   Chapter Six: Dying for a Shed

114 Chapter Seven: Chinese Calamari

134 Chapter Eight: Nostalgic for Now

146 Chapter Nine: A Turnip for the Books

159 Chapter Ten: The Wild West (Coast)

178 Chapter Eleven: A Lego Adventure

195 Chapter Twelve: Idiots Assemble

221 Chapter Thirteen: A Very Brown Christmas

232 The Last Splash

235 Acknowledgements

Introduction

When the shit hits the fan, you win or you learn.

It's a charming expression; one I have heard a few times, as the last forty-two years have etched their progress onto my skin. I've come to regard it as an unbreakable truth; one of the fundamentals of the universe.

I've always had a deep-seated belief, a kind of mantra: 'Dream. Believe. Achieve.' I like it so much I had it tattooed on my left wrist. It's there to remind me – in those moments when everything goes to hell – that the bigger picture *will* be brighter. And I really believe that. Not in a learned way. I didn't read a self-help book or take life-coaching lessons. I haven't joined a cult and it's not something I trained myself to think. It's more a kind of ingrained faith; one I only realised I had a few years ago. You might think I'm naïve, and fair enough, but I have mostly assumed that if I just keep going, things will turn out well when I'm 'all grown up'. Luckily, so far, they have …

So, why *Code Brown*? In the course of writing and researching this book, I realised I had stolen the phrase from the world of nursing. It was unintentional, I promise. I would never knowingly steal from nurses. They do an amazing job, under very difficult circumstances – which too frequently involve me – and, more importantly, I'm scared of most of the nurses I know. The phrase 'code brown' probably means exactly what you think it does – that someone (or something) has shat themselves. For nurses, dealing with that is an every-day human reality. In my line of work, and the day-to-day of trimming hooves, it's thankfully a lot less human in nature. But it is a more frequent hazard of the job.

Like a darker – or browner – version of the song, 'Into Each Life A Little Rain Must Fall', literal and metaphorical code browns are everywhere. For me, they are learning experiences. I truly believe they make us who we are. Like thousands of feet on sandstone steps, they shape and hone our characters over time, and this book is full of mine.

But before we dive into the madness – the storm chasing, the runaway vehicles, the stranded sheep – I want to start by telling you where I'm at now. Not just geographically, but mentally, emotionally. The real challenges rarely start with a bang. They heat up, slowly, like the contents of a pressure cooker. So, here's a bit of context. If you're new to my world, my name's Graeme, and I'm a professional cow-hoof trimmer with a YouTube channel, a lot of energy, and a life that some-times makes more sense in hindsight than it does in real time.

I live in Wigtownshire, on the western side of Galloway in the south-west corner of Scotland. It's a rough and tumble kind of place, full of character and characters. I feel like I know at least half the population within a ten-mile radius of my home. We're so far from anywhere in Scotland that our nearest city is Belfast. It's a lot like an island. Word gets around in a community like this. Tales of screw-ups travel farther and faster than just about anything else, and they come back home with a liberal dose of ridicule.

When I was younger, I had a different attitude. When something unexpected happened, if someone or something caused hassle, or hurt or humiliation, I'd think, 'of course it did', or 'just my luck', or, more likely, 'fucking typical'. I suppose these kinds of things happen a lot when you're young. Now in my older, hopefully *slightly* wiser years, I have come to see those episodes as positives. I like a challenge and I'm always on the hunt for an opportunity to learn. In these moments – when it might be easy to give up because of sheer hassle – you have the chance to find out who you really are, to push through, build your resilience and, hopefully, better yourself.

I have another tattoo on the back of my left hand. I like tattoos. I got my first one, a Chinese dragon, when I was eighteen. I can pretty much chart my life's progress by the sentiments I've had inked on my body. The tat on my left hand says, 'do more', and it's there for occasions like these. It's a reminder to keep going when I feel tired or disheartened, when I've lost my appetite for the daily battles. It's that extra kick up the arse, just when I need it.

If I can only do more than the other guy, if I keep going when he gives in because he's tired and he can't be arsed, I have a competitive edge. Again and again, I find myself digging in, keeping on going, because I glanced down, saw those words, and just did more. A lot of people say hard work beats talent. I wouldn't like to call it either way, but I suppose, with my tattoo, I'm unconsciously buying into that whole idea.

Let's be brutally honest though. Listening to me isn't always a great idea. I have been wide of the mark on MANY occasions, and that's what this book is about. *Code Brown* details those in-over-my-head moments, the muck-ups and fuckups, the disasters and full-throttle code browns, where victory hasn't always been the outcome. If humility is a thing we grow into, the quicker you're able to embrace your own blind spots and limitations and work around them, the quicker you'll grow in life.

From filling my boxer shorts with the previous night's Guinness and curry, while travelling on a crowded train hurtling through the Yorkshire Dales, to telling a national radio audience about my neighbour's penchant for breakfasting on testicles, to literally blowing myself up, this is my calendar of code browns, my personal almanac of teachable-moment shitstorms. This book is a window into my world; a snapshot of the trials and tribulations of a Scotsman working on the harsh west coast of his native land, through the seasons of the farming year.

And of course, because I'm me, I tend to rack up more than one code brown a month, plus January gets two whole chapters because it feels like it's about two months long.

You win or you learn. It's not so much a motto, as a way of life. Welcome to my world!

Chapter One:

Sound Bites

It was the third of January, and I was excited. I was driving along an icy, single-track road, through the Galloway Hills, in a car that may yet prove too powerful – a Lamborghini Urus Performante, if you're interested, and a big Italian SUV if you're not. It's a car designed by a man who seems to have been obsessed with hexagons; a family wagon capable of doing a hundred and ninety miles an hour, flat out. Right then, I was doing about seventeen. I was all keyed up, and not just because I was trying to avoid ditches and keep everything hexagonal. I was buzzing for what I was about to do. I didn't need to leave the house that day. I could have sat on my mobile or the laptop. I could have phoned it in. But the more my life revolves around social media, the more I crave real life and the up-close-and-personal. I was about to go on the radio. That's not a first, now. I've even been on TV a few times, but the novelty has never quite worn off.

It was supposed to be a quiet start to the year. Our house revolves around family. Ashley's one of six kids. I'm one of five, and we love to have everyone round. I'm happy in the kitchen, cooking and experimenting. When we renovated our house, we paid the most attention to the kitchen. We squeezed in a whacking great island, somewhere the kids can sit and eat breakfast, do their homework – usually under protest – and talk about their day. It's the hub of the home; the place friends sit and chat. We added two ovens for the big cooking sessions, and we kept everything bright to capture the light, and to boost my mood. I suffer from Rapid Cycling Bipolar Disorder, so health hacks are a must, and light is a good one.

My culinary obsession goes back to my days managing restaurants and gastropubs, but I don't get to do much cooking now. I love entertaining but last year was a big one, and cooking for the whole tribe, timing everything so it comes out of the oven at just the right time, and keeping the booze flowing, takes a bit of doing. We had it easy this time. We spent Christmas and Hogmanay in other people's houses. New Year's Day was going to be a chilled one.

'Are you still going to the gym?' Ashley said, shortly after I hauled myself out of bed.

'Yeah. Might as well start the year off properly,' I groaned. I managed three gym visits every week last year. Some weeks I was away, and some I was busy, but I made up the sessions I missed and averaged it out. Lifting weights gives me mental clarity. It makes me feel stronger in my body and mind. I feel better when I'm not testing the limits of my T-shirt in all the wrong places. Consistency is probably easier when the gym is just down the hall. I put in glass doors, mainly to bring in more light, but there is another advantage. I can see the weights and feel the guilt, every time I pass. Even with that view, you can never underestimate the potential of good intentions going bad, so I try not to go it alone. Mark, Ashley's brother, and Derek, who is married to Nicole, Ashley's sister, come round on Mondays, Wednesdays and Fridays, every week. A bit of healthy competition – some might say mild bullying and bitter rivalry – help keep us all motivated.

Everyone calls Derek 'Yorkie'. I realise that makes him sound like a Yorkshire terrier, but he's really the opposite of that. The anti-terrier. He's more like a bear; a big, broad, fluffy, guy whose shoulders need to be substantial just to hold the weight of his head. He gets his name from his dad, who used to drive trucks. Between the seventies and the nineties, the UK confectioner Rowntrees – and then Nestlé, who bought Rowntrees – advertised their chunky Yorkie chocolate bars as a snack for 'manly' men, usually truck drivers. Life is like that round here. Memories are long and you can inherit a nickname your dad had, or your dad's name, even if isn't on your birth certificate. Or you get named for something you did when you were a kid, and it's yours for ever. I know of a mild-mannered man called 'Killer' who earned his nickname from swatting a wasp one day.

Derek might be the most competitive person I know. That's why training with him works. I'm keen to compete and we push each other forward. Unfortunately, he is quite a bit stronger than me. He bench-presses my fifty-kilogram dumbbells without thinking about it but at least that gives me something to aim for.

Every year we host a big family barbecue at the house.

There's food and drink and a bouncy castle, for the kids – and sometimes the adults, too. A couple of years back, we had a school sports-day theme. There was an egg-and-spoon race and a three-legged race. Yorkie annihilated the competition in the men's sprint. Mark told everyone it was because he thought there was a Chinese takeaway behind the dyke[1] at the end of the lawn. That's where Mark fits into the gym equation. He loves to wind everyone up. Mark's wiry, as we say in Scotland – thin and hardy. His arms are ridiculously strong. Just don't tell him I said that. I showed him my drone, one Boxing Day. I was flying it around the street we lived on at the time, in the village of Kirkcowan.

'Are you not worried about hitting something,' Mark said.

'No. Not at all. It's got sensors that stop it flying into anything.' I demonstrated, by flying the drone directly at my house, and promptly smashing it into the pebble-dashed wall on my neighbour's house. My aim was a bit off, due to the amount of drink we'd consumed. I'd also forgotten I'd put it in sports mode, and that it overrides any crash-prevention measures. We watched as the drone clattered to the ground.

A wide grin spread across Mark's face. 'Well, that'll be that then,' he said. Straight to the point.

As much as anything, these gym sessions are a good excuse to hang out and catch up. There are days I don't feel like working out and I'm tempted to cancel. Then the boys arrive,

---

1. A dyke is a dry-stone wall in Scotland. In many other parts of the world the word is used to describe a channel or a ditch. We like to be different!

we lift some weights, and I feel good. Mark arrives smelling of cow shit, literally bringing his work into the gym with him. He's been on the same farm since he was thirteen. I'm pretty sure he's indispensable now, like part of a machine you forget is there until it goes 'pop', sending everything haywire. He lives and breathes farming. Derek runs the local branch of Tarff Valley, an agricultural depot that sells everything from animal feed to welly boots. He drives around the whole area during the week, picking up all the gossip. He can tell you who's selling this and doing that and diversifying the other. Between the three of us, we have a lot to talk about, in between winding each other up, obviously.

So, I decided to head into the gym, and Derek and Mark decided they might as well join me. Nicole thought she might as well come for a drink, so Mary, Ashley's mum, and Grant, Ashley's brother, arrived for drinks too. In the end there were seven adults and about twelve kids in the house.

And that, right there, is our life, in one perfect snapshot: a chaotic mess, bursting with family and energy and laughter.

I used to make cocktails all the time. I took a real pride in my mixology. I still like to mix my drinks properly, so Ashley now refuses to make them herself. I was back and forth, between the gym, making the girls porn star martinis, in between sets on the weights. I love playing barman, but I'm a bit conflicted. When I'm busy being the host I miss half the party, but when I'm not busy I get bored. In the end, everyone else got pissed but I had to drive to Glasgow early the next morning.

The email had come a couple of weeks earlier. 'Can I ring you for an interview?' was all it said. A one-line message. I didn't really look at it. I didn't recognise the sender at first. It looked a wee bit suspicious. We get some strange messages. Usually, the sender is trying to sell me something or get me to promote something, or more often, just outright defraud me. Ashley gets the job of answering the emails. She intercepted this one, saying she was really sorry, but that we don't respond to interview requests without having prior contact with the interviewer. She was polite, but firm. She shut the sender down, assuming he was just another scammer. Three or four minutes passed. And then I noticed the email address, in particular, the @bbc.co.uk part at the end.

John Beattie hosts the morning show on Radio Scotland and wanted to know if we could do a phone interview. I told him we could do a *proper* interview. I didn't mind driving to Glasgow. In fact, I would prefer that. Honestly? I fancied a drive, and a nosy round the BBC Scotland studios. I find any situation like that interesting. I'm on all the major social media channels, so I'm used to seeing my mug onscreen. In 2025, YouTube is second only to the BBC in terms of UK viewing figures, but the processes of 'real' TV and radio production still fascinate me. I love the frenetic energy. I love the monitors and microphones, the 'on-air' light – that's still a thing – and the runners buzzing in and out asking questions. It's busy, like you're at the throbbing heart of things. I really love the idea of being a radio DJ, and talking to an audience live, from a studio, with all those buttons. I'll take any chance to experience some of that buzz. Being there in person almost always makes for a better conversation. You build up a rapport you just don't get from Zoom or phone calls. I read somewhere that over 50% of communication is non-verbal, and I can believe it.

My interview was scheduled for eleven. I survived the drive and rolled up at BBC Scotland's vast glass headquarters at Pacific Quay on the Clyde, early. The building was quiet. The second of January is a public holiday in Scotland, a hangover – pun intended – from the days when Christmas was banned following the Protestant Reformation. Most people had taken the third as a bonus day off this year because it was a Friday. It might sound odd, given where I live, but this is my favourite view in the world: the Clyde waterfront with its glassy buildings, the asymmetric curve of the Clyde Arc bridge, and the armadillo skin of the SECC in the background. I love it so much I have a painting of it in my hall.

I arrived sharp because I thought I'd better give the producers time to tell me what to do, but I needn't have. There's none of that. I sailed through security and into the studio, where they sat me on a couch.

'We'll call you in a minute,' one of the runners told me.

I sat listening in to the show I was about to be on. That's enough to make a man nervous. 'We're live in fifteen seconds,' another runner said. And then I was on the air, talking to John Beattie.

There were screens everywhere, maybe twelve in total. The studio itself was bigger than I expected and lit in a purply-blue glow. There was no view to the outside, but I didn't miss it, there in the reassuring odour of 'warm computer'. John stood behind his desk. He, too, was more imposing than I had imagined. He used to play rugby for Scotland, so I suppose he should be. He had a laptop and a pile of notes. I sat down, opposite him, and I couldn't help feeling like I was back in the headmaster's office. There seemed to have been a change of plan. They told me Kamalyn Kaur, the psychologist, would be on after me, but John introduced her up front. She was on the line. This was happening, there and then. It occurred to me that the whole phone-in concept made more sense; that we would have been on different lines. Then, another thought jumped into my head. Three months before this, I released my first book, *Bruised Sole*. I wrote about my life and my struggles with Rapid Cycling Bipolar Disorder and ADHD. I opened myself up. Now I wondered if this was some kind of ambush. It was all happening so quickly, it felt haphazard. They'd even spelled my name 'Graham'. Was I in trouble? And if so, what kind of trouble?

I said hello to Caitlin, the producer, as she passed. She shot me a smile as she disappeared behind a glass wall, into a room full of more buttons.

'Do you want a photograph?' John said.

'Ehm ... Yeah, okay,' I said.

I got the feeling this was standard practice, so I stood next to him. I realised I'd probably end up on social media, and that the jumper I had on was a bit loud. It's an exaggerated multi-coloured houndstooth check. Funnily enough, I didn't think many people would be seeing me that day, on the radio.

The interview started out well. John asked me a few questions about how it all began. I told him I swore I'd never become a hoof trimmer, that I thought it was the most disgusting job in the world but now I love it. I told him I'd wanted to create videos for my customers, just as an extra to educate them on hoof health, but that my audience hadn't been the one I'd expected at all. When I'd checked my demographics, the viewers were in the inner cities of the United States, and not farmers from Scotland.

Kamalyn came on the line to explain why people might get addicted to my videos. John told me *he* was addicted to my videos and that's why I was there. I was flattered, and interested in the theory, but it was a lot to take in. *Exhale.* Kamalyn said my videos feed a very human need; a deep-seated drive to create order from chaos. She told us that trimming hooves, shaping them and healing them, and helping cows, provides a kind of resolution, and that humans are hard-wired to seek out that feeling. Kamalyn said the effect of watching me fix hooves is similar to the sensation the viewers would get if they fixed them themselves. It's the same reason people like to watch garden clean-ups or tidying videos.

Things felt like they were going well, but my brain wouldn't slow down. The voice in my head told me not to lower my guard. I worried about farming-related questions, as I always do. It feels like everyone's going after farmers these days, and that anybody reporting on the industry has an agenda. I'm constantly trying to defend farming families and their values, but I want to do it with positivity and openness. We have nothing to hide, and I'm happy to talk about anything. I want to show how farmers look after their cows, but sometimes it's a balancing act. I wanted to sound excited and stay true to what's important – helping the cows – but I also needed to talk to John about the journey of the Hoof GP because that's why I was there. If more people watch the channel, they can help get the word out. The wider world will learn what really happens on farms, rather than the propaganda people are fed on a daily basis.

I tried not to mention the book I had just released, but John asked me about Fiona the Sheep and her rescue and whether I'd been offered TV opportunities. I told him I had, but they'd always been fluffy concepts, like driving across America, visiting small farms, that kind of thing. I'd had the same kinds of approach with book offers, but I knew I could only start writing when it mattered, when I found something real and raw. It turned out *that* was my life. *Bruised Sole* matters to me. Mental health matters to me. Talking about both felt natural, but even as we did, I regretted it. It felt forced, like I was plugging myself. *Ground, swallow me up*.

That wouldn't be the first time I'd said the wrong thing on the radio. I have form. In my early twenties, I managed a restaurant. The Windmill Tavern was, as it sounds, a tavern in an old windmill. It was in a place called The Fylde, a square of coastal plain in Lancashire, in the north-west of England. The Windmill was part of a brewery chain, but you'd never have known it. Along with Danny Brannan, the head chef, I worked hard to make the place a success. We liked to challenge our-selves. We smoked our own duck in a repurposed oil-drum. We engineered strange concoctions, none of which were on the official menu. I had some tough conversations with the bosses, but when we'd finished, we had an award-winning gastropub. I was proud of what we did there. Looking back, I'm happy to say I knew how special that was at the time. Our success got us noticed, and that's how we got the call from BBC Radio Lancashire. They spoke to us by phone and asked for our professional take on a few things. I didn't think much of it at the time, but a few weeks passed and I got another call. They were doing a feature on sweetbreads, and they wanted to talk to me.

A lot of people think sweetbreads are lambs' testes. I know. People actually eat those! I believe they're called prairie oysters in some parts of the world. Lovely. Sweetbreads are, in fact, the pancreas or thymus of a lamb or a calf. Neither of those sounded particularly appetising to me but we were running a gastropub. It was our job to push things. Any publicity from the biggest radio station in the area was good, and the more memorable the better. So, when the producers asked if there was any way we could prepare some sweetbreads – and yes, they did mean lambs' bollocks – and bring them in, I was very happy to oblige. After all, Danny would be doing the work.

Danny sourced the goods and cooked them up. Then, we drove our precious cargo to Manchester, in my Vauxhall Corsa. If you're reading this in Australia, that's a Holden Barina. If you're reading it in America, GM don't sell cars that small, so just imagine something the size of a new Mini, but a whole lot less cool. It's very easy to underestimate the volume of traffic around Manchester. That's how we came to be sitting in a tailback listening to radio updates about how we would be on the radio soon, and then the updates about us *not* being on the radio soon. My stress levels were through the roof. I'd scored valuable free publicity for the pub, and now I'd blown it, because I was bloody late. We knew we'd missed the show. I think we got a message telling us not to bother coming in, but we ignored that and ploughed on anyway.

BBC Radio Lancashire's studio in the mid-2000s was not like BBC Radio Scotland's in 2025. It was dark and musty, like a school or a library that has seen better days. Everything was painted magnolia. They had the tech, but the atmosphere could only be described as 'institutional-dank'. The station staff told us they'd talk to us after the show and record what they needed. They were interviewing Danny, so I sat in the background, watching, trying to keep out of mischief. Danny talked them through what he'd done with the feast in front of us. He had done his best to prepare everything delicately. He was, and is, a professional, but you know there's only so much you can do with lambs' balls. He'd prepared them along with some actual sweetbreads, and he tasted them himself. He talked about them for a while. I'll never forget how he cut into one of the offending items and yellow, creamy gunk oozed from the middle. Mouth-watering, and not in a good way.

We were with the DJ, but off-air. I was relaxed, and up for talking. When Danny's interview was done, there was an awkward pause. It might not have been that awkward in reality, but to me it was a gaping conversational void. It had to be filled.

'You know in Scotland people will literally cut them out of lambs and put them straight into the frying pan and eat them?' I said. 'Actually, I used to work on a farm when I was a kid, and the farmer would cut the testicles off the lambs and pull them out. Then, he'd select his favourites and take them into the house. His wife would cook them up and he'd eat them. Within fifteen minutes, the balls could go from the lamb's nut-sack into his belly.'

The DJ sat in front of me, mouth agape. I decided I'd probably said enough. Then, he started to laugh.

Filled with renewed confidence, I started talking again. 'He would come back out and you could really smell it on his breath.'

I swear to all that's holy, I didn't think they were still recording. I don't think *they* knew they were still recording. But they were, and guess which part of the interview BBC Radio Lancashire played the most of? Yup. There wasn't too much of Danny talking about his cooking, but there was a hell of a lot of me detailing the sight of this imaginary farmer with the mangled remains of lambs' family jewels stuck between his teeth. Business *did* pick up at the Windmill, so I suppose my efforts paid off. We never did put Danny's dish on the menu though. Thankfully.

I'd love to say that I learn from everything; that I moved on from that episode and vowed never to be caught out in the same way again, but I soon found out that my Radio Scotland interview had all been filmed and then released online. I hadn't said anything too daft this time, but I really wish I'd worn another jumper.

Kamalyn's professional view solidified something for me though. Sometimes, in the middle of editing a video, I'll think, 'cut to the chase Graeme'. Later on, some viewers will comment the same thing, but I know the story leading up to the payoff matters, just as much as the moment of revelation itself. The slow parts of the trim are soothing, almost hyp- notic for some people, even when there's nothing major to fix. I'm told it can help relieve anxiety. I didn't discover that on purpose, but I did *have* a purpose – fixing cows' feet, and sometimes, that initial drive is all you need.

The alternative reasons people watch do influence what I'm doing, now. I try hard to film from the right angle, so that people can appreciate the form of the foot and get the best view of the clean-up process; the scooping everything out and cleaning it all up. To me, that is what's satisfying in real life. Transferring that into a video is the challenge. I'm determined to get in closer, to make things as real as possible for the viewers.

If I could invent smell-o-vision for my videos, I would. But for now, I'm just relieved that my radio interview skills have improved since my days at the Windmill, even if my choice of attire should probably have come with a loudness warning.

Chapter Two:
The Calm Before Wotsisname

January was strange this year. The temperature hovered around eleven or twelve degrees centigrade for a lot of the month. We didn't see so much of the cold I love at this time of year. We had a week where it dropped as low as minus six, and the low-slung, amber sun showed Scotland at its best. I love layering up and heading off to work when it's freezing outside. The cold, hard metal of the crush and the spray of hoof chips tortures numbing fingers.

Before I became a hoof trimmer, I freeze branded cattle. Freeze branding is the cleaner, less stressful equivalent to old-style hot branding. That's illegal here in the UK. We use extreme cold, instead of extreme heat. The branding irons are chilled to minus eighty degrees centigrade to kill the pigmentation in the cow's hair follicles. The hair grows in colourless, and the number comes up white. There are no absolutes, of course. Weirdly, if you freeze brand a Suffolk sheep, the number grows in black.

When I was freeze branding, I would leave my irons in a dry ice and alcohol soup overnight. Everyone else I knew tipped theirs out. Profit margins were never good in freeze branding, so I did anything to get a head start; to keep the irons cold and move quicker. I still have scars on the bit of skin between my thumb and forefinger, from the burning cold of that metal. Those scars get tender when temperatures drop, but nothing warms you like hard work.

On those cooler days, we bring the cows in, and you can see their breath in the morning sun. I could spend a lot of time just trying to capture that with my camera, but I have to resist. This time of year brings so many surprises. Migrating Canada geese land on our lawn, their honking calls echoing out into the surrounding silence. We're off to work before the sun comes up, and the purple and red and blue sunrises on Wigtown Bay would just about break your heart.

I don't know why, but I've noticed cows produce a lot less manure in winter. Maybe it's something to do with their diet, or the cold, or a combination of both, but it's a real thing, I promise. I've monitored their toilet habits closely, odd as that must sound. The upside is that the crush and everything else we use is a lot easier to clean at this time of year.

These days we have Cameraman Graeme with us two or three days a week, filming the action. When he's not out and about on farms, he's in the comfort of the team office, putting it all together. Graeme is fantastic at what he does, but he has an interesting take on cows and the outdoor life in general.

His favourite catchphrase of the moment is, 'I honestly have never felt this cold in my life, ever'. At least, I think that's what he's saying. Anyone who has ever seen Graeme in one of the videos will know he has quite a strong accent, even by Galloway standards.

Years ago, while working on a fishing boat, he tore the tendons in his left hand on a winch. That led to nine operations and a long-term disability. That kind of thing is common round here, in an economy based on farming, fishing and forestry. I know farmers who have lost fingers and even limbs just trying to clear jammed machines. Graeme's injury means that he feels the cold more than most. His hand is super sensitive. That's partly why he wanted to become an editor and avoid freezing days in the great outdoors. He started out as 'Editor Graeme' but slowly morphed into our cameraman. Now we're constantly telling him not to do stuff, to take it easy, but he likes to get stuck in. A few weeks ago, I overheard a commotion at the front of the crush. He was getting frustrated with a cow.

'What's wrong?' I said. 'Is it the cold?'

'I cannae get her thumb in,' Graeme said.

That made me laugh, a lot, but I knew what he meant. Cows have small claws on the backs of their lower legs. They're called dew claws, and they're kind of primeval thumbs. Like a lot of ruminants, cows have four digits: two on each hoof and two dew claws on their fetlocks. Graeme was raising this cow's foot for trimming but struggling to get her dew claw into the right position in the cradle that holds the foot. Craig and I were still laughing when I had to concede he was right.

The first time Graeme came out with us, he asked if he could go and film some cows because he liked the way the light was hitting them. I knew I'd made a good hiring decision then. You can't teach that kind of eye for detail. That said, I may have created a monster. He regularly takes the piss out of me for the dodgy editing in my early videos. I tell him we didn't all have the privilege of learning from a master.

The other day, someone asked if he was Croatian. I think it was the accent. When he first started, I told him not to try to talk 'properly'. Even when he did, it didn't come out clearly. I told him I'd put subtitles in front of him if I had to, but I love the authenticity. Graeme has a local accent. When I was younger, I hated my Wigtownshire accent. I've been penalised for it when going for jobs (more on that later). For a long time, I tried my best to sound generic, like I was from nowhere. Now, after living away, and picking up hints of Dumfries and Lancashire, I miss my old voice. I hear the romance in the lilting Galloway Irish and I'm jealous. I've learned to talk slowly in my videos so people can understand me, but every so often the camera catches me talking to a farmer at full speed. I sound very different, and I like it, so I leave it in the final cut. You can see the effect in the comments.

Graeme was editing a video a few weeks back and he shouted across to me and Craig. 'Come here a minute. I just need to ask you something.' The three of us gathered round the desk and he played a clip back. On the screen, Graeme was talking, rapidly. After about thirty seconds, he paused the video and eyed both of us for a reaction. 'Be honest. Have you got any fucking idea what I'm saying or not? Coz, I don't even know, like.'

'Nope. Welcome to our world!'

He thinks so far ahead of his mouth, he has to speak like that to keep up. His accent is pure class, and though it's a running joke, I'd much rather caption what he says and have him talking, unfiltered, onscreen, than have him speaking in some kind of Shire version of received pronunciation. What would that even sound like anyway? Combined with his dry sense of humour, his accent is hilarious, and the dynamic at work can be interesting.

I'll say something cheeky to Graeme, and without cracking a smile he'll say something drier back, deadpan. One of us will laugh eventually, but it's not a common occurrence. Meanwhile, Craig will be getting twitchy. Craig is a good egg. He wants everyone to feel comfortable. He really doesn't want anyone to get the wrong idea or take offence. So, Graeme and I will be trading mean jokes, and Craig will explain what each one meant to the other. And no one will be laughing – just to add to the comedy effect of it all. Working with the pair of them *does* make me laugh, all day though, even if it's mostly on the inside.

Craig is married to my little sister, Susan. As I write this, the pair of them have gone off to Bergamo, chasing the elusive winter sun. Work is strange. Craig is supposed to start work at half past seven each morning, but he turns up at quarter past and organises everything to death. He checks the equipment. He double checks that everything is charged, then loads up all the tools and the gates we need, stocks up on blocks, bandages and my very own Hoof Grip Pro glue and fires up the computer recording system for the day. When he is done, we are ready to roll.

When we get to the farm, we work in unison, without having to think about it. I cover whatever he's not doing and vice versa. We don't even need to talk. I mean we do, because there's far too much fun to be had giving each other a hard time, but it isn't strictly necessary. Craig does things the same way I do. Our approach is consistent, because we've trained in the same place and worked together for so many years. He does things exactly as I ask him to, and that's important when the stakes are so high. In the end, it's all about the cows. He's a fantastic hoof trimmer, and he makes it all look easy. The only slightly tricky bit is when I have to get him to do something different.

Craig doesn't 'do' change. He'll 'um' and 'ah' and pull faces and he'll give it a go, but he won't like it. Our crush came with paddles to support the underside of the cows' feet, and in theory it looks good, but the laws of physics have other ideas. Newton's Third Law states that 'every reaction has an equal and opposite reaction'. I started to notice feet bouncing on the paddles when I was using the grinder to pare back the hoof horn. The movement makes the trim unpredictable. A jumping hoof, and a powerful grinder with a sharp, whirring, bladed disc, could make a real mess of a cow's hoof.

We could rectify that to an extent, but I'd rather not have to. The design of my KVK crush means the feet are already supported at the top. One day, I decided we'd try trimming without the paddles. They're in a downward position until we pull them up with a lever, so I just stopped raising them. I left the foot supported at the top, but floating in the air with nothing to bounce on. I tried that approach myself for a week, and the bouncing went away. Then, it was time to tell His Craigness. He wasn't keen. He was used to the paddles. I wasn't really watching him but every time I did look around, I'd see him try a wee bit the new way, then give in and raise the paddle. Then, he stopped even trying that little bit.

'Come on. We need to try to do this,' I said. 'I really don't want to take a chunk out the cow's foot. It's a good solution. You'll get used to it.'

'Right. Okay.'

He tried it for a day. The following day, we resumed our merry dance. Every time I checked, he'd give it his token effort and then he'd put the paddle back up. Eventually, I snapped the lever that raises the paddle clean off.

'You didn't have to do that!' Craig said.

'Well, I can't fix it now,' I said.

He's quite happy not using the paddle now, because he has no choice. But when he's on holiday, I feel like I've lost an organ – my spleen or my pancreas – something I didn't quite know the importance of until it goes AWOL.

The twenty-fifth of January is Burns Night. It's also Susan's birthday, so it was always a busy time in our house when I was a kid. My own kids become busy spouting off Burns' poetry, getting ready for recitals at school. Rabbie Burns is Scotland's national bard. He's such a big part of our identity, he gets his own day. The only other people who get their names on days round here are Jesus and St Andrew, and they're not even Scottish.

Keir did the traditional 'Address to a Haggis' and he absolutely loved it. I'm not sure if we did when we were younger. My older brother James did, though. I remember him getting a book of Burns poetry in his stocking one Christmas and reciting all two hundred and twenty-four lines of 'Tam O' Shanter' for us at three in the morning. The 'Address to a Haggis' is long enough at forty-eight lines, but it certainly helps to pass the time as the seemingly endless dark days drag on and on.

January went out with a bang this year. The weather was mild to begin with. We then got a seasonal flutter of snow, so I went to find the kids. 'Right. Sledges,' I announced, in as serious a way as I could. I loaded up my Raptor and we drove deep into the Galloway hills, along a track surrounded by pine trees. There was a light dusting of the white stuff everywhere but nowhere was properly deep. We drove on until we found a couple of little hills and I pushed the boys on their plastic sledges. Inevitably, a cartoon lightbulb magically appeared above my head. The Raptor is a big, murdered out, black pick-up on swollen BF Goodrich tires. It's grippy. We had sledges with string on them. There really was only one thing to do.

I tied the sledges' pull-strings to the back of the truck, and the boys took their places. We coasted along at six or seven miles an hour. Kier and Campbell were staggered, one behind the other, to avoid any side-on collisions. Campbell wanted to go faster but Keir was holding on for dear life. Up ahead, I could see a big corner, so I pushed on a little bit, just enough to let the boys drift round the bend. Keir didn't love that, so we stopped, and he climbed back up into the cab. Campbell told me he was staying on his sledge. I fired the truck up again and we carried on along the track. We approached another corner, and I could see a fallen tree, a big one, blocking the road. I brought the Raptor to a halt and jumped out to take a look. The tree was still around the bend, but Campbell couldn't quite see it because he was sitting at an angle.

'Campbell, there's a tree up ahead,' I said.

Campbell looked disappointed.

'Okay. You'll have to hold on really tight when we go over it. I'll speed right up, so I can jump over, but when you hit it, you might go high in the air, so make sure you hold on. As tight as you can, right?'

'Okay,' Campbell said, seriously. He stiffened in his seat and gripped the sides of the sledge firmly.

I climbed back into the truck.

'Let me out!' Keir said.

'Don't worry,' I said. Behind us, Campbell was in the sledge, happy that he was about to jump over a bloody huge tree. Or so he thought. I drove as far as the outer branches, then I pulled up and jumped out, laughing.

Campbell looked past me, beyond the pick-up and the massive tree. 'What's wrong, is there something else there too?' he said.

I looked at him, wide-eyed. 'What did you think was gonna

happen? Did you really think I was gonna fly you over that massive tree?'

'I don't know but you said it was okay, so I thought ... hold on.'

Nowadays, the Met Office – the UK's weather forecasting service – works alongside its Irish and Dutch counterparts, Met Éireann and KNMI, to name storms. That's a new thing here. We didn't have named storms until November 2015. We would see US hurricanes with names on the news, and that did make them sound more ominous. I think the Met Office was taking note, just itching to do the same thing, and all that power might just have gone to their heads. Or maybe our storms just got bigger. They'd been promising us a huge storm for a week. I hadn't been expecting their predictions to come to much. I was wrong.

With the three countries involved, the pre-approved names are a mixture of English, Irish and Dutch. It had taken me six or seven attempts to pronounce Storm Eowyn's name, and I still wasn't sure. There's a hill at the back of Laigh Kirkland. It banks down behind the house and when the wind comes from the east, we get hammered. Luckily, that doesn't happen a lot. We're more likely to get it from the west. Eowyn started whipping things up during the night. We could hear things getting wild, but we didn't think much of it. In the morning, when we left the house, there was an immense din, the weirdest, most menacing boom I've ever heard, like driving over a level crossing on the railway. The wind wasn't powerful down by the house, but it was blasting the hill behind us and deflecting up into the air. In practical terms it was just blowy, but that noise was incredible.

Ashley was worried about her papa. She'd spoken to him three times on the phone but he's ninety-two years old and she wasn't happy, so we decided we'd drive the eight miles south, down the Machars peninsula to where he lives in Garlieston. I'll be honest. I just wanted to drive somewhere because it was a stormy day and I was fascinated, like when I was a child.

Our power went out early. Ashley's mum is only a quarter of a mile away, in Wigtown, but she had electricity and a wood-burning stove, so we took the boys there, then went to Garlieston. I'd never ever seen as many downed trees. There were twelve or thirteen of them in the space of five miles. The farmers were there, chopping up trees and pushing them out of the road with load-alls, as the storm raged on. Ashley's papa was fine. We knew he would be because Ashley had spoken to him THREE TIMES that day, but it's always best to be sure. We sat and chatted to him for an hour. He worked on the railways for years and he's always telling me stories from that time and when he was stationed in Egypt during the Second World War. He's small and pudgy and always full of cheer. He loves corduroy and cups of tea. He's the heart of Ashley's family and she calls him 'the best man in the world'. I try not to take that too personally.

The pick-up took a pounding on the way home. It sits high but it's low-sided, and I was surprised how much of an effect the wind had. We stopped off at Gordon Agricultural. It's a place I've known all my life, as a farmer's son. It sells machines of all shapes and sizes, so I bought a generator and I back-wired it into a plug socket and got power back on in the house. The normal supply ended up out for a long five hours, but we were sorted now. Ashley started the fryer – even though I asked her not to – and it blew a fuse, but I was secretly happy because it proved the setup was safe.

By the following morning, the power was back on. Job done. I congratulated myself and decided we no longer needed a generator, so I drove it out to my best mate Dave's house. The police were on the road as I drove down. They were working with the farmers, now, cutting up and moving more trees. The verges were covered in limbless trunks, severed branches and leaf litter. It's a productive mess. In the city, nobody would do this. It would be a job for the council, who would call in professionals to do it at inflated prices. It would take time. There would be health and safety assessments. They'd close off the road. Here, there could be dozens of trees down and they'd all be cleared by the following day.

Dave's power had gone out. Like me, he's a farmer's son, so he likes to get his hands dirty and get things done. His parents moved the family up from Cornwall in 1989, so his accent is stuck somewhere between here and there. When we were much younger and my granny was still alive, she liked to ask him questions just so she could hear him talk.

We got the generator up and running. It was noisy but it was in a shed, a good sixty metres from his house, so he couldn't hear it. Later, I was in the gym with Yorkie, filming our workout for a vlog. I told him I'd done a good deed, but that Ashley thought I'd been a bit quick off the mark.

'What happens if you're out of electricity?' Yorkie said. He barely finished his sentence. On the word 'electricity', the lights went out. *Typical.*

Ashley appeared at the glass door. It was dark, but I could tell there was another storm brewing. Derek was there, so her tone was light-hearted, with a hint of absolute fury detectable only to me. We carried on with our workout, improvising with a couple of torches.

'I can't believe you're just doing your workout,' Ashley said. 'Go and get the generator from Dave.'

'There's no way I'm going round to Dave's and disappointing his kids and disappointing him,' I said.

We were having a more in-depth 'discussion' about it as the power came back on. Ashley was ecstatic and I captured it on video.

'Obviously the power was only being switched off to fix something or test something so job's a good 'un,' I said.

It went off for twenty minutes so that was a safe assumption, surely? I didn't need to get the generator from Dave. We were still laughing about it when the power went off again. I told myself it'd come back on. We waited a few hours, and it didn't come back on. I told myself there was no way I was going to Dave's. We went back to Ashley's mum's house and stayed there until eleven, enjoying the heat of the fire and cutting-edge appliances, like light bulbs. We went home, to sleep in the nineteenth century.

'Look, if it's still off tomorrow, I will tell Dave that I need the generator back,' I said.

The next morning, I deliberately slept in until ten. There was no power. At half past ten, I phoned Dave and said I needed to come and get the generator. I felt like the worst person in the world, like I was coming to take away their happiness. The kids were looking at me with their little faces. Susan, Dave's wife, was busy in the kitchen and didn't speak to me at first.

'You really pissed off at me or just busy?' I said, living dangerously. On my way home, I still wasn't sure, and as I pulled into the drive, I saw the outside lights, already on. I very nearly turned and went back. The only thing that stopped me was that I felt ridiculous.

Overall, the damage was fairly insignificant where we are. We get big storms at least once a year. They clear out all the dead wood and anything that's even slightly precarious, so when the next storm comes through, a lot of the bad stuff is long gone. The electricity is a different story. A week later, a couple of the local substations were still out.

This was billed as 'a once in a generation' storm and it didn't disappoint. They measured the wind at a hundred and thirteen miles an hour at the Mull of Galloway – the end of the hammer on the end of Scotland. It was ninety-four near Dave's house and it peeled one of his shed roofs off like the top of a sardine can. The power was immense. I'm very aware of the dangers it poses, but I love extreme weather, and I love being out in it. As children, we'd go up to the second hole on the golf course on the shore at St Medan. We'd stand at the back of the green, on an edge at the top of a steep drop down to the fourth green, and we'd put our arms out. We'd lean into the wind like human kites. The aim was to get to 45°, and if you pushed it that little bit too far, you'd stumble forward and over the edge. I've always loved stuff like that, and I hope I always will.

One Christmas Eve, twenty-odd years ago, when we lived in a house on the harbour in Port William – an eighteenth-century fishing village, ten miles from Wigtown, on the shore of Luce Bay – a storm came in and wiped out one side of the main road. It washed great chunks of tarmac into the Irish Sea and threw up boulders from the beach, in exchange. It was terrifying and glorious all at once. We were out for dinner that night – Mum, my stepdad Davie, Susan and me. Mum was so worried when we drove home along the coast, that she gave me a St Christopher she'd been saving for a present. Thinking about that hits me a little bit. I'm still sad I lost that St Christopher at school.

Years later, Ashley and I were saved by the timing of another storm. The Trading Post was an old warehouse and shop down on the harbour, in Garlieston. It's long since been demolished but there are old storage sheds down there – warehouses built from sandstone and local whinstone. They carry the scars of the years gone by and they are beautiful. One side banks onto the beach and the other onto the harbour. The building could be made into a stunning restaurant, and that's exactly what we wanted to do. We had everything in place to buy it. We'd agreed the price. All that was left was to transfer the money. And then there was a storm, a massive storm. The wind wasn't as strong as Eowyn, but the tide was high, and the waves rolled in like thunder. When everything was calm again, we went down there for a look. Rocks the size of footballs lay everywhere – up on the harbour and all through the building. The windows were gone. We decided it was a sign, and that we shouldn't buy the building. We both imagined how we'd feel if we had restored it. It would have been destroyed. That storm was a saving grace, a warning shot, but it would've made an incredible restaurant.

Today we have more construction plans. We recently bought a nearby farmstead and we're in the process of converting it into houses for holiday letting. While we were waiting for the builders to come, I was like a small child counting the days until his birthday. We're using local tradesmen for the build, and I've been trying to scavenge as much granite as possible from all the farms around me. I want to use it for the garden walls and the entrance. There's a huge pile of granite at The Knock, and a granite ball that sits on stacked stone at the end of a dyke. It's an interesting feature. Mum and Dad built it that way, probably fifty years ago. I'd love to have something like that. I'd love to incorporate the stone from home, somehow. To think Dad probably lifted this bit of granite would be an awesome thing.

More than anything, I've been desperate to start paying for it, so I don't spend the money on more cars, but at least we now know how to weatherproof the farmstead against the ravages of storms with unpronounceable names.

Chapter Three:
Big Boy Boots

My heart was trying to beat its way out of my chest when we arrived. I forced the car door open and pushed myself into the light. I placed one foot in front of the other and counted every step. It was February 2024, and I had been asked to attend the University of Glasgow's Veterinary School to give an introductory talk in front of about three hundred students. The talk was on cattle healthcare, and as part of it, they'd asked me to do a demonstration with cadavers.

Fourteen years previously, I had no intention of ever going near a cow's hoof. Now I was about to deliver a lecture on them, and when it came to animal husbandry, I couldn't think of a more prestigious location. It's bizarre, but there I was, talking to lecturers and vets, talking to three hundred students, on a stage, on the outskirts of Glasgow. You might think, 'but you're on YouTube, you'll have loads of confidence'. Honestly? I was shitting myself. I'd nearly said no. Then, the voice in my head told me this was good for me, that I needed to broaden my horizons, and push myself. That's why I was there. That, and I felt it would be good to have a couple of hundred future vets on my side whenever the inevitable terrible thing happened, and people started thinking badly of me. I get very few negative comments online. When they do happen, I can deal with it, because I know my job. On some level, though, I think I'll always fear the looming spectre of cancellation. On the bad days, everyone hates me. Everyone thinks, 'he's up himself', 'he thinks he's the man', or 'he's got all this confidence'. Really, I'm just a frightened man, hoping no one finds me out.

'I don't think you should do it,' Ashley had said, when I told her. 'You're putting yourself in the firing line.'

Ashley knows I put myself through the ringer, every time I get up there. She's inevitably the one talking me down when I'm a fight-or-flight mess. I wanted to prove to myself that I could do this, that I had the technical chops to teach these students, to give them something useful. I *needed* to do this, but I was terrified someone would challenge me or call me out for doing what is very nearly a medical procedure, when I wasn't a qualified vet.

I had overprepared. I knew what I was going to say. I had fired questions at myself, relentlessly, pre-empting the ways this could all go wrong, but within five minutes of wandering the campus, I realised I didn't even know where to go. I walked the halls looking for some signs. People approached me with smiling faces and hellos. It's getting pretty normal, all this, wondering if people are looking at me because they think I'm the Hoof GP, or if they're even looking at me at all. Normal, but still weird. You don't want to assume *anything*. You don't want to look arrogant, but you don't want to seem like you're overthinking it. Some days these approaches catch me off guard. The surprise jolts me, and leaves me sweaty palmed, riddled with questions about what just went on. Afterwards, I worry about what I said. How I came across. Was I too much? Or too dismissive? Was I a disappointment?

So, there I was, at the veterinary school, bricking it. How had this happened? I regularly trim at a dairy farm down at Baldoon, just outside Wigtown. Baldoon is a fascinating place. There's a castle with a well-known ghost, and a Second World War airfield. Ruth works at the farm I go to, but she's also a veterinary student and she is always curious about hoof trimming. So much so, she suggested me as a speaker for a symposium the veterinary school was holding on hoof care. I know her fairly well, so I was comfortable about turning up, until I did, and realised she wasn't there.

I found my way to the right place and introduced myself to the event organisers. They told me I'd be doing the demonstration with two vets. Up till then, I'd had it in my head that I'd be presenting alone. That's what I'd prepared for. Now they were throwing me into a different mix. I'm not naturally an organised person, but when it comes to presenting, or talking on stage, I like to stay on my toes. I want to have more than I need up my sleeve, because it simplifies everything. It means I don't need to think about anything other than what I'm talking about.

I love teaching. I love passing on what I know. That's what my YouTube channel is all about. That's why I started making videos and why I love having a team around me. But as rewarding as it is, I don't naturally *want* to do it. I don't know if I'm any good at it. I have to force myself out there, but once I'm talking, the barrier falls, and when I get into the flow, I breathe a massive sigh of relief. Normal service is resumed. I'm immersed in the conversation. Every time I go on stage, I tell the audience it *is* a conversation. An open dialogue. 'Ask questions. Please! Ask questions.'

The students at the veterinary school did not let me down. They were engaged, hungry for knowledge, and I remembered, then, that I *wanted* to be challenged. It helps push me forward, keeps my knowledge on the boil. I never want to be caught off guard, so, no matter what I'm doing, I always want to know everything I possibly can about it.

There were a lot of dead cows' feet. We were on concrete, in big pens, made from fifty-mil tubular steel. The kind of metal that would hold back a rhinoceros. Maybe that's a possibility in a vet school ... A single bare bulb hung over my head. It was functional. This place looked a lot like an abattoir, or what I imagine an abattoir would look like. I got to work, using one of the cadavers to demonstrate how I trim a hoof. I shaped it slowly, with my knife. I cut the toe to the correct length with a pair of nippers. The feet were clamped in loosely, so I used the grinder sparingly, to avoid the hoof flopping and jumping with the pressure and movement from the whirring blades – a look that wouldn't inspire confidence. It'd be a long time before the students got near a grinder anyway, so this approach was more relevant. The grinder was only there to show how a hoof trimmer works at speed in the real world, where efficiency matters.

When my demonstration was done and I was happy, I let the students have a go. They were keen to get stuck in, and I walked the room, enjoying their enthusiasm. I graded their efforts and handed out titbits of advice. At first, I wasn't sure if most of these people wondered who I was. But the more conversations I had, the more people told me they were watching my videos.

I used to have to explain what I did when I met someone new. 'Hoof trimmer' is an unusual occupation. My stepdad, Davie, used to tell people he was a cattle chiropodist. He delighted in telling them he'd got out of a speeding fine once, because the police couldn't spell 'cattle chiropodist'. Hopefully it was just the second bit they struggled with. More and more, I start to explain what I do, and people say, 'Oh yeah, I've seen that.' A lot of them call me Graeme in a familiar way, as if they've known me for years. It can be unsettling, but it's endearing, not because you feel proud that they know who you are, but because it's always a positive interaction.

The session was invigorating. When we left, my head was up, and my shoulders were back. I was surfing a dopamine wave. I was, for once, content; happy I said yes to doing it. I may have been mildly terrified, but the result was good. It's a process I seem to have gone through my whole life.

I used to get nervous before job interviews – so nervous, a lot of times I just didn't go. I screwed up vast periods of my life because I couldn't do the thing I needed to, like go to college to hand in the project I had worked on or go to class and explain *why* I hadn't done the thing I was supposed to. Then I read something; I can't remember where. It said you shouldn't think about the destination, you should just concentrate on putting one foot in front of the other, on all your steps linking up, and before you know it, you've arrived at your destination anyway. Concentrate on the journey, on getting through that, and there's no time to worry about the destination. Maybe I need that tattooed somewhere.

Talking at the biggest hoof-trimming conference in the Unites States was a big challenge. We were in Orlando a year ago, at the Hoof Trimmers Association conference, in the Drury Plaza – a huge hotel, right in the heart of Walt Disney World. Talking in front of these hoof trimmers is probably the most daunting thing I've ever done. I was the keynote speaker, and that came with some hype. The kind of hype you need to live up to. Then, the absolute worst happened. I ran through some checks beforehand. I asked if everything was organised, and the equipment would work well. I phoned ahead. I emailed ahead. I was very clear. I needed to know it was all planned properly, and that nothing could go wrong … in theory.

Everything was set up and ready to go. And then I walked on stage, and nothing worked. The TV screens were blank. The mic was dead. I just stood there, with a few hundred people staring right back at me, and I wondered how many of them were expecting a punchline. If you'd asked me, I'd have expected to go into meltdown; to start sweating, to wet myself. Weirdly though, I didn't panic. I knew I'd done the work. I was prepared, so I didn't worry about it. The equipment came back on a few moments later, and the talk went remarkably well.

Then came the part I had been dreading and looking forward to in equal parts. Ashley, her mum, my mum and the boys had all come out to Florida with us. We treated them to a villa and a holiday, stateside. Then, Ashley got ill. That meant she couldn't come to the conference. We'd decided to make the most of the trip and do a meet-and-greet with fans from Florida. Meet-and-greets are unpredictable. You can never be sure whether two people or a thousand people will turn up. Pretty much everyone from the world of hoof care was represented in one room. There were vendors, suppliers, people associated with the industry, and there was a meet-and-greet with yours truly, the Hoof GP. I had pictured a queue of one. I decided I would feel like an idiot, but it would be over quickly, and that might be better than a crowd. We had a green stand, promoting my green Hoof Grip Pro glue, and I prepared myself for no one being there. Then, they started to arrive, in dribs and drabs, a little earlier than planned. And then *more* people started to arrive. Eventually, I looked out at a queue of a hundred people or more. I felt my stomach tighten, and I ran for an exit.

I snuck out to the fire escape and stood at the back, with a pint of orange punch Steve Wunderlich had given me. Steve's a hoof trimmer and farmer from Pennsylvania, with the best name in the world and a hardcore attitude. One look at his Facebook page and the picture of him in a black and white, cow-patterned suit should tell you everything you need to know. I sat there drinking Steve's wunder-brew and I phoned Ashley.

I felt tears pooling in the corners of my eyes. My vision began to blur. 'I can't do this,' I said. 'I can't do this. I *can't* do this.' In my mind's eye, I stood in front of the other hoof trimmers, and I looked stupid. I heard the awful things they must surely be thinking. I knew I wasn't enough.

'You're gonna have to put your big boy boots on,' Ashley said. 'Just get back in there and go for it. People have made the effort to come and see you. You can't let them down.'

I knew she was right, but that didn't make it easier. I dried my eyes on my shirt sleeve, then I turned, and pushed my way back through the fire escape.

When I got back to the booth, there were six or seven hundred people standing in line. They had come out from all over Florida and Tennessee and the neighbouring states. People from every walk of life: Black, white, Asian, Latin American, farmers and city slickers alike. Mum was waiting. She wasn't supposed to be doing anything, but she was in for a shock. I informed her that she was now the queue manager. Somebody would come forward to speak to me, and Mum would speak to the next people in line. At first, they didn't realise she was my mum, but the news soon rippled backwards. Mum owns a coffee shop. She is a people person. She was in her element. There were so many people coming up with Hoof GP merch, hoodies, T-shirts and caps to sign. She ended up getting her photograph taken with a few of the fans and someone got her to sign something. I can't remember what, but they got my mum to sign it, which was just cool.

We were halfway through the day when a man stepped

forward. I can't remember his name. I wish I could. He was about my age, my height, and athletic. He was shaking.

'You know, I'm so close to tears, I nearly turned back multiple times,' the man said.

'What do you mean?'

The guy said he'd driven for four hours and at every junction on the interstate he'd tried to turn off, but he'd fought the urge. 'It's like, no! I'm not! I'm going to make it. But I was so nervous about coming …'

I started to laugh.

'Well, it's not funny,' the man said.

'No,' I said. 'I'm laughing at me. I was outside on the fire escape, about half an hour ago, in tears because I didn't want to come in here. I was so nervous. So, you're exactly the same as me.'

It was a special moment, meeting all those fans, but it was bloody terrifying. I had to force myself to get out there and do it. And to have this man tell me he felt the same way, and for me to be comforting him, and him to be comforting me, just seemed right, somehow. The whole occasion, meeting all those people who had gone out of their way, was massively humbling. One girl had queued at a bakery in Disney Springs – a shopping centre in Orlando – for two hours just to get us some muffins and cookies. I thought, *that can't be true*, but she told me the name of the place and when we were in Disney Springs a couple of days later there was an almighty queue for the same bakery.

Ashley and I love country music. It reminds me of home

in a lot of ways. It's as American as a cheeseburger, but there's a direct connection through bluegrass and the Scottish and Irish folk music taken across the pond by generations of settlers. To me, country is a rose-hued romantic take on rural life, and it gets me right in my soul. We couldn't resist going to Nashville. Nashville is something else. Live music is everywhere. You walk down Broadway, and a different sound drifts out of every bar. Everyone seems up for a good time. We never saw any trouble, and no one seemed to be drunk. That might sound like an odd thing to say but coming from Scotland it feels rare.

Tootsie's Bar was crammed full of people. The music was good, and the drinks were flowing. I turned to look at Ashley. She looked a bit pale.

'What's wrong with you, are you still feeling ill,' I said.

Ashley tipped her head, indicating a spot somewhere behind her. 'That woman's got a photo of me.'

I squinted over her shoulder, to see a lady going wild. She held her phone at arm's length, with a picture of Ashley onscreen. People usually notice Ashley before they recognise me. Outside work, I'm not covered in cow shit.

We went to the Johnny Cash Museum. Ashley's a big fan and we had high expectations, but there wasn't much to it. Later, we saw a show at the Grand Ole Opry. I've never been anywhere like that. I don't suppose there *is* anywhere like that. We sat in the massive auditorium, on benches like church pews. None of the crowd sang, like they would here in Scotland. When the acts performed, it was to a hushed, reverent silence, and we felt the music. The Opry works in a different sort of way. They don't tell you who's performing when you book your ticket, and the artists don't do traditional sets. There might be eight or nine acts on the bill. Each of them does four songs. It's like the musical equivalent of a taster menu in a swanky restaurant. There's an element of risk, but we were lucky, and we hit it just right. My favourite singer, Lainey Wilson, topped the bill that night, and she blew me away. She sang the songs I love like I'd never heard them.

Jamey Johnson was announced, and the place went wild. This stocky guy with the biggest beard I've ever seen walked out onto the stage. There was no bullshit with him, no bravado, just a dignified walk to the microphone. Then he started singing 'Lead Me Home', a song I'd never heard, and I could feel it vibrating through my whole body. I've never been anywhere where music has affected me like that. I started crying halfway through 'Lead Me Home'. It's about a guy at the end of his days, about to meet his maker. It hit me hard, and it was a very special thing to be able to share with both of our mums. Between singers, I mentioned to Ashley that one of the artists was amazing, and a voice from behind us said, 'she was amazing, but not as amazing as your videos!' I turned to see a smiling Hoof GP fan, just to add a surreal touch to the evening.

Later, we were in a bar, and an old cowboy approached us. He grabbed Mum by the hand and just took her off square dancing. Mum was like a schoolgirl. She jumped up and got stuck in – to the dancing I mean, not the cowboy. She was in her element. When I was younger and we went to Cyprus and Ayia Napa, there was a feeling of everyone being desperate to look cool, but in Nashville there's none of that. Everyone – from nineteen to ninety – is there to enjoy themselves, and they're all dressed differently. I never used to feel like I fitted in. In Nashville, you don't have to. *And* they have cornbread! We'll revisit in a few years, when the boys are older. Then, we can take full advantage of the festivities.

It was good to get back home to Scotland, though. February is one of my favourite times for walking in the woods. The snowdrops are surfacing, and the forests are yawning back into life. The trees won't be green for a while, but the ground becomes carpeted in crocuses. The weather is grim as I write this. It's two degrees, and the damp cold seeps into your bones. You wash up the crush at the end of a farm visit, your hands drenched in freezing water, and your fingers burn with every surface they hit.

Work is hard in this humid cold, so we try to get away somewhere warm and sunny. This year, we spent some of our February in the Maldives. When we got married, Ashley wanted a honeymoon. I've written about this before, but we couldn't afford a holiday then. It was either get married, with no honeymoon, or just don't get married at all. At the time I told Ashley that if I could ever afford to take her to the Maldives, I would. Fast forward a couple of years into married life and we were completely broke; skint, in debt. I promised, then, that after ten years of marriage we would renew our vows, and if we could afford to renew our vows, we would do it on a beach, in the Maldives. We're now lucky enough to be able to do that, albeit a few years later than I'd hoped.

I'll be honest. When it came to it, when we finally achieved what felt like a fantasy, when we were booking the flights and packing our cases, I realised I wasn't excited. I kept thinking, *We're going to a tiny island. I'll miss work. I'll miss home. I'll get bored.* I dreaded the journey. We were only going for seven days, and it's a long way. We boarded a British Airways flight at Heathrow, on a wet, windy night. One of the air stewards looked at me as I passed. 'I just started your book yesterday,' he said. 'You've gone through a lot of things I've gone through.' I looked at him more closely. His name badge said 'Glenn'. He was stocky with close-cropped dark hair, but I couldn't picture him on a farm.

I smiled and thanked him, then I wandered through to my seat and settled in. Ten years ago, I couldn't have afforded the flight, let alone the holiday. Now I was *on* that flight, and someone was there to acknowledge the change. It felt poetic, hyper-real. Later in the flight, Glenn came and spoke to me about his life. He said his dad and his brother watched the channel too. He told me how he'd worked on dairy farms but wanted to become a flight attendant. I love it when people go against the grain and follow their dreams. There can't be too many people who have taken Glenn's exact career path.

When we arrived in the Maldives, I was apprehensive. I knew we were supposed to have a butler, and I couldn't stop thinking about it. My instinctive reaction when I heard was, 'my God no! We'll just refuse that when we get there.' The idea of somebody helping you is great, but the idea of having a butler just isn't my bag. It feels too much, unless his name is Alfred and he comes with access to a cave full of gadgetry and vehicles. His name was Yusuf, and he came with a villa and a golf cart, but it didn't take me long to realise Yusuf loved his job. That made having him around very easy.

We stayed in a villa on the shoreline, next to a wooden boardwalk in the shape of a lemon. The interior was open plan, all wood and white curtains, with a massive bed at the heart. The island is four hundred metres of white sand in an azure blue stretch of Indian ocean. They brushed the beaches every morning and night, just so there were no footsteps in the sand. It was pristinely satisfying, like real life ASMR.

Yusuf told us there were six restaurants. Food is my main hobby in life, and each restaurant is a five-star dining experience. *Jackpot.* My mouth watered at the thought. We were all set to dive in when Keir started throwing up. He told us he didn't think he was seasick, and between us we decided it must be food poisoning. Somehow, Campbell, Ashley and I were fine. Keir threw up seven or eight times on the first day. It was a strange kind of illness. He seemed stuck in a cycle. He'd vomit and then he was fine for a while. Then, he'd be sick again, and so on. I felt terrible for him. He kept apologising, saying he'd 'ruined' the holiday, but by the following morning, he'd done his best Lazarus impression and got straight onto the slide that ran from the roof of the villa into a sea as warm as a bath.

We were on a reef, and you could stick your head under the water and see fish of every colour. Most holidays don't live up to the professional photos on the websites and in the brochures, but this time, the photos could never have done it justice. We swam round the island, exploring the reef, spotting sharks, jellyfish and a sea cucumber, which is the most disgusting looking thing I've ever seen. Naturally, the boys picked it up.

On the seventh of February, at the tail end of the holiday of a lifetime, we stood on the beach as the waves kissed the shore and we renewed our commitment to each other, with vows we had written ourselves. I had spent four days putting my vows together, in between snorkelling sessions with the boys.

I can't keep a secret. Ashley detests secrets, and our lives are intertwined anyway, to the point that all of my apps are linked to all of her apps and all of my bank accounts are linked to her accounts. Even Ashley's phone is linked to mine. Keeping secrets is nigh on impossible. But I thought it was only right that I give her a new ring. The problem was that if I bought her anything she would see the money coming out of my bank account, the confirmation email landing in my email account, etcetera. So, I had to be a wee bit sneaky. I asked Mum if I could use her debit card to buy the ring. A week before this holiday, I stole one of Ashley's other rings, to make sure I had the right size. Within an hour of me taking it, she noticed, and we had a massive argument. Ashley kept saying I'd touched it last. She was – of course – quite right, but I kept telling myself, *No, don't give in. Don't give in.*

Later, I went to Glasgow to get some work done on a tattoo on my upper arm. The new ink featured a woman's face with

Ashley's name hidden in the design. When I was done for the afternoon, I had half an hour to run to James Porter, one of the top jewellers in the city. All of Ashley's rings centre around diamonds. She doesn't like coloured stones, but I wanted to get her something different. I picked a yellow diamond, surrounded by white diamonds, on a platinum ring. I asked the lady in the shop if it was possible to exchange the ring. 'Yeah,' she said. 'Do you mean to get it resized?'

'No,' I said. 'To choose a different ring. There's no way she'll like the one the I pick. No matter how much I like it.'

I think it's stunning. I had managed to buy it, and sneak it out to the Maldives, undetected. But, when we arrived there, we realised there was a jewellery shop. Ashley soon spotted it, and she stopped to stare through the glass at a blue, sapphire ring, with diamonds. 'That's incredible,' she said.

'Yeah, it's beautiful,' I said. 'But … have you seen those yellow ones?'

'Oh God,' Ashley said. 'I would never wear anything yellow on my hands.'

'Yeah,' I said. 'But these ones aren't that nice. I think you get some beautiful yellow diamonds.' Ashley looked at green stones, blue ones, and at more white diamonds, and I kept trying to bring it back to the yellow ones.

'No. Honestly,' she said. 'I don't know why you keep going on about them. I think they're disgusting.'

All I could think was, *Great, she'll absolutely hate it*. But as we stood on the beach, renewing our vows, and I handed her the yellow stone, I could see she was blown away. She absolutely loved it. Just this once, I seem to have picked a piece of jewellery my wife likes.

My other concern was that it would feel weird, just the four of us, in a ceremony, on the beach. I needn't have worried. I feel like we're the four Musketeers, getting through everything together. Some people like to get away from their kids for a couple of days, and I get that. Family life can be hard, but we've never really felt that way. One day is about our maximum. After that, we start to miss them too much. I know. We're sad, but it means they have to come everywhere with us. Our dream was always to have the kids be part of our ceremony. We weren't completely alone. There were eight or nine other people on the beach. There were drummers and singers, a celebrant and other people keeping things running smoothly, but it felt like it was just the four of us. This was especially apparent when we were presented with a three-tier wedding cake, and I remembered that Keir and Campbell hate the stuff. We gave it to the staff, hoping someone might eat it.

Later, as night fell, the resort team transformed the altar into a dinner table for us. We sat there, lit by candles and moonlight, under a blanket of stars, on a sandy beach. The waves lapped at the table and the kids played in the sea. The food was amazing, and country music played in the background. It was a special, special time. One I know I'll never forget.

When we got back, I was apprehensive. This was Craigie Boy's first time being left in charge. Part of me imagined coming back to a world of chaos and broken things, but Craig had managed not to break anything, which was impressive. The holiday was amazing, but you pay for it afterwards. There is always a mountain to climb when I get back. Farmers know when I'm off, so they give Craigus Minimus an easy time, mainly because I ask them to. When I got back, I had a double helping of cows. I arrived home at 8 a.m. and I was back on a farm for half past eight. I was – as they say round here – absolutely hanging out my bum. That holiday was worth it though.

The dynamic duo filmed some things while I was away, but had to deal with more wind, and horrific audio as a result of that. I edited it all together with voiceover. In the footage, Cameraman Graeme says, 'Every time I take a cow out the pen there's less in there.' Craig takes the bait and explains that it's *because* he's taken a cow out of the pen.

I once listened to them having a forty-minute conversation about a 'moose' in Canada because 'mouse' is pronounced 'moose' in Scotland. I don't think either had a clue what the other was talking about, and at some point, for some reason, Craig decided Graeme was talking about koalas. I might have added a bit of fuel to the fire, but it was fun listening to them talk about the size of a moose in Canada, realising they were both right and both wrong at the same time.

Later in the day, we filmed a podcast episode for the Herd – my Hoof GP YouTube Members – and Graeme wondered aloud if any plastic surgeons out there would sponsor him with a hair transplant. As far as I can see he still has a full head of hair, but I decided not to point that out to him. Back to reality for us!

Chapter Four:

Spring is on the March

All through the winter we're desperate for spring. Every year, I fool myself into thinking the beginning of March is the point it'll kick in, that if we can just get through February, the world will be all right again. No matter how pretty it can be around here, the conditions are still harsh to work in, and, as always, we're not quite there yet. This is when cows' feet are at their worst. It has been wet for months. Scotland is damp year-round, of course, but the months up to March are a different kind of wet. Most of the cows we see are inside and they've been there for six months, subjected to slurry and standing water. All that moisture and the concrete are a catalyst for all the little problems that can build up over time. Beef cattle are notoriously bad. They're out in the fields more, and away from concrete, but we don't see them for extended periods. Then, all of a sudden, it's March, the fields have turned to bogs, and the toll shows on their feet. It can be challenging, but it's never dull. We're working on different cases. I don't like to see animals suffering but I enjoy putting my skills to work. It keeps me out of trouble.

By March, the days are stretching and we're working harder. It takes five or six times as long to trim a lame cow's feet. Mentally, that's a slog. Progress is slow, and it can put you on a downer. Dairy farmers see their cows every single day and hoof health impacts directly on their productivity. Lower milk production is a big red flag. Beef farmers don't necessarily see their cows daily. They live more natural lives, on grass. That's the best thing for them, but nothing's perfect. A lot of people think that the biggest issue with foot care is concrete and manure, and both are problematic, but the job of fixing them is complicated most by our wet climate.

If you walked around in your stocking soles on dry concrete, your feet would callous and become hard, but if you walked around in water, they'd eventually start to crack, and you'd get a nasty case of trench foot. I like to think of it as the difference between cake and biscuits. If you leave a cupcake and some shortbread exposed to the air, the dry biscuits will go moist, and the cake will turn dry – unless you leave them in my house and I solve the problem by eating the lot. Osmosis forces water molecule concentration to equalise across the membranes in living cells and other permeable materials. Hoof horn is like the biscuit, but it's in contact with the ground, which at this time of year is a lot wetter than the air. Hoof horn is organic. It might be tough but it's permeable. It gets saturated. Water is important for life. Living things thrive on it, including bacteria, and bacteria flourishes in cows' feet this time of year, so we're dealing with some more crazy cases.

A few years ago, we were on a farm over towards Kirkcudbright: Scotland's artists' town. It was a particularly hard shift. When we're trimming, feet are dirty. Normally, we're in a shed because the cows have been brought inside, but this farm didn't have any sheds. We were out in the middle of a huge field, in an area the Royal Air Force use as a bombing range. We herded in a hundred and fifty or so cows, with calves 'at foot', as they say round here. We had to separate the mothers from their young calves. Cows and calves were mooing their heads off. It was a busy, busy atmosphere, thick with noise and the smell of freshly turned soil. It was also dangerous.

Most people worry about bulls, but the most aggressive animal you'll find on a farm is a heifer – a cow who hasn't yet given birth for the first time – who is heavily in calf, or a cow who has just given birth. Every year, we hear of someone being killed because they ran or cycled through a field of heifers, thinking they were safe because there were no bulls. I've had a few run-ins of my own.

When I was nine or ten, we had a field full of Limousin cross Angus heifers, and we were trying to bring them in. They were close to calving, so it would have been around March-time. We were moving them, and like all cornered animals, they wanted a way out. I was halfway up the field. The heifers were nearly at the gate. One of my brothers was below them, down the hill, and somebody else was above them, blocking off a wide gulley we used to store scrap metal. The heifers looked around, weighed up their options, and concluded I was the smallest person there. That made me the easiest way out. They turned as one beefy mass and they ran. I stood frozen, watching as they gathered speed. This was really happening, and I could do nothing. I was a small boy in the path of forty, hormonal, black and red heifers galloping straight at me. I couldn't stop them, so I ran. I can still feel the fear as I write this. The rush of blood, and the absolute certainty that they were about to stampede over the top of me.

They were getting closer. Then, above the noise of the swooshing wind and the thundering of a hundred and sixty hooves, I heard my dad's voice.

'Stop and turn around! STOP AND TURN AROUND!'

I stopped, I turned, and I threw my hands up over my head. The heifers parted and flowed past me, like I was a rock in a stream. I remember thinking then that I needed to trust the people around me, and not just my dad. I realised in that moment that if I hadn't committed, if I hadn't put my trust in him, I would have fallen under those hooves, and that would have been the end of me. Looking back now, as a father myself, I realise Dad must have been absolutely shitting himself. He knew what I had to do but he couldn't get near me. He knew I had to do it for myself. I couldn't go near cows for a couple of weeks after that.

Another time, I actually got hit. There was a Hereford cross cow, black with white feet, in a cluster of whin bushes. I was only just a bit older by then, but keen to be useful. I decided I'd better check what was going on with her. I poked my head into the whins. I could see she had just calved. She still had cleanings – what we call the afterbirth – hanging out of her back end, and I wanted to make sure she was okay. My brother James was with me. The cow probably just meant to warn me. James later said he realised something was wrong when he saw me shooting out of the bushes on the head of an angry cow. I was somewhat reluctantly straddling her. Then she just stopped and calmly went back to her calf. It was like being hit by a bus, then standing up and thinking, *Did that just happen?*

Growing up on The Knock, one of our two family farms, and getting into scrapes like that, earned me an intuition, an almost complete understanding of what a cow is thinking. I say 'almost' because you get unpredictable cows, in the same way you get unpredictable people. But you *know* they're unpredictable, and you make allowances, like when Craig hasn't had his morning coffee. There are some things you can teach. The vast majority of cows prefer to turn left, for example, and, given the choice, they like to walk uphill. I suppose you would too, if you had all that strain on your knees.

A cow's depth perception isn't the same as ours though. On farms, a lot of livestock handling areas, races, chutes and pens are made up of steel gates. Usually, the gates are formed from tubular bars, so you can see straight through the gaps. That perspective works for us, but it confuses the hell out of cows. They're fine when they're close up and they can see or feel those bars, but add some distance and they focus on the gaps. They can see through to the other side. They don't understand why they can't go there, so they'll give it a try. If you put solid sides on a race, the cows can't see beyond that, so they keep on walking, following the line of the wall you've created. In a similar way, if you have a cow in a massive pen and you want her to go down a tight race, the best way you can help the cow is to walk right round the edge. That shows her what's going on, spatially.

A lot of people think cows are calmer in a crush because they're restrained, but that's not the case. Some hoof trimmers use a squeeze crush. The sides of these move in, holding the cow, and that calms them. Temple Grandin famously designed a lot of cattle handling equipment based on cow psychology. Temple is autistic and designed a 'hug machine' or 'squeeze box' to calm herself when she's feeling anxious. Swaddling a baby has a similar effect. The squeeze crushes don't stop cows moving because they can't move, but because they feel secure and they're calmer.

Cow psychology is fascinating to me and understanding it can make a world of difference when you're handling them. But there are some things you can't teach. Some things have to be learned through hard-won experience.

Anyway, we were on this farm outside Kirkcudbright, bringing all the cows and calves in to separate them out, looking for lame cows, and overgrown feet. We herded them into a big corral the farmer had built using movable hurdles. It was a horrendous day. The grass was soaking. Craigasaurus Rex was with me, and he was pissed off with the rain and the claggy[2] ground – all pitted and bubbled with hoof prints. There were too many cows to get through, and when we'd separated them from their calves, the noise was horrendous,

---

2. 'Claggy' is a commonly used word in the UK, meaning, damp, sticky or muggy.

the calves shouting for their mums and the mums were shouting for their calves. We couldn't get all the calves out before we started, so we had to separate the escapees in the race and return them to the corral. I was enjoying myself. I liked the challenge of the conditions. It took me back. I felt more like a farmer that day than I had done in a long time.

We saw some horrific feet – massively overgrown claws that hadn't been trimmed in twelve months. Hoof horn grows around ninety millimetres in a year. If cows are on soft ground their hooves can't wear down. In an ideal situation the growth and wear are equal, and feet stay nice and dry. You'll see that on harder land in warmer climates in parts of North America, South America and the UAE. In Scotland, the feet just keep growing.

I walked round the crush, slipping and sliding all over the place, quietly proud of the conditions I was powering through. I relished the idea of the rectangle of untouched grass when we moved the crush away, and that we'd be able to see our tracks.

The March weather may be hard on hooves, but it can be equally brutal on the humans. There's a farm we visit where they milk Jerseys. We can spend long days there, trimming a hundred and twenty or a hundred and thirty cows at a time. There's a rough road that goes straight through the middle of the property; a dirt track the cows use to get to and from the milking parlour twice a day. It's a linear bog of muck and manure, the kind of thing I might go down in the pick-up, but not if I have the choice.

One March day, Craig and I were up there, working in the pelting rain. It was horrible. We had a small, tin shed roof to protect us from the worst of it. It's what we'd call a 'New Zealand-style' farm. The cows graze out in the open all year. Jerseys are small. They're light on the land, and they can stand up to a bit of weather. We were trimming away and a DPD van appeared. DPD is a courier firm in the UK, known for its red- and white-liveried Mercedes Sprinter vans. This particular van drove down the little dirt track while Craig and I watched, wondering what we'd missed. Maybe there was a cottage down there we didn't know about … maybe the driver wasn't just wheel-spinning his way through the muck and mire for no reason. As he slid past us, we agreed that either he knew where he was going, or his satnav was playing tricks on him, and either way, we'd find out in a few minutes.

A good half hour passed before the white nose of the van reappeared round the bend. Unfortunately, in that half hour the farmer had come along and moved some of his fence gates so that his cows could be guided back to fresh pastures – some of his *electric* fence gates. Craig and I watched the DPD van fishtailing back along the track. The driver came to the first gate, stopped and got out. He was a big guy, the type that obviously spends a lot of time in the gym. Top-heavy, in shorts and a red fleece, he tiptoed through the mud, towards the fence.

I looked at Craig. Craig looked at me. I remember thinking, *No way*, as we both realised what was about to happen. The man reached the wire that barred the road. He snatched it out of the way. There was the briefest pause, then he shot backwards as the electric pulse kicked in. We could hear the squelch as he landed on his backside, in cow shit. He slid and jerked back to his feet. Craig and I tried very hard not to laugh. Thankfully he was okay.

The courier stomped his way back to the van, climbed aboard and drove on to the next gate. He tiptoed through the muck, still covered in a layer of the stuff, as though the track was consuming him. We watched him take in the white ribbon in front of him. Presumably happy it couldn't be electrified, he lifted it, then his arm and his head recoiled again. He probably realised at that point that electric fence technology has advanced a lot, and not all wires look the same.

We could hear him shouting and swearing, now, covered in crap and going ballistic. He got back into his van and put his foot down. As he was about to pass us and we were trying to conceal our giggles, he slammed on the brakes and slid to a messy halt. He was fifty yards from the crush and walking in our direction. Craig decided he needed to 'look for something' in the shed. There was no way I could keep a straight face. I took in the mountain of roid-rage marching towards me and decided I might be about to die.

'I'm so sorry, mate,' I said. 'That was the funniest thing I've ever seen in my life.'

Luckily for me, the giant had a sense of humour. He took it on the chin and asked for directions.

It's not just rain we have to contend with in March. These days we have a massive shed, and everything is inside. That avoids a lot of precipitation-related hassle. We have a concrete road, and if there's ice, I can remove it and salt it to keep it clear. Nowadays, the worst aspect of the weather is having to go out in it. In 2012 that wasn't the case. Back then, Ashley and I rented a barn conversion, outside a place called Creetown, just across the bay from where we live now. The house was amazing, with local whinstone walls and wooden beams, but it was at the bottom of a hill, down a rough, gravel road.

In 2012 we had a big snowstorm in Scotland. I mean it's all relative. This is Scotland so it wouldn't have been the kind of big snowstorms that the USA or Canada gets, for example, but it was pretty big for here. On open ground, in the centre of the fields, there was about six inches of snow. I can hear some of you Canadians laughing now, but that's a lot for us. The wind did its worst where there were structures. It drifted the snow, covering cars and walls, piling it up on roads. We're not used to that, and we're really not set up to deal with it. I emerged from the house as it started to snow, like a hopeful kid.

Ashley is a hairdresser to trade. At the time she was working in a salon in Newton Stewart, about seven miles away. I pictured her getting snowed in. In my head this scenario did not play out well. I dug out my phone. 'I think you're gonna need to come home,' I said, when I got hold of her.

Ashley didn't want to drive. The roads were already getting slippy and she didn't fancy dancing on ice in an Audi TT. This was my chance to come to the rescue. 'Don't worry, I'll come and drive your car back,' I said, as heroically as possible. Then, I layered up and started walking.

I know what you're thinking, and you're right. I wanted to go out and play in the extreme weather. Of course I did! I trudged my way to the old railway line. The rails themselves are long gone, like many in the UK, but sixty years ago, the towns here were connected, all the way down the Machars Peninsula, and west, to Stranraer and Portpatrick. They used to call it The Paddy Line. It was the main route for people and goods to catch the ferry to Ireland. As I crunched on, the snow grew heavier. It fell in great clods, cocooning the world in a white, fluffy coat. I was almost at Newton Stewart when Ashley phoned. Our next-door neighbour, Jack, had given her a lift home. I had phoned him earlier, knowing he was working in Newton, and he'd come up trumps. The problem now was that the snow had piled up. Now *I* needed rescuing.

I phoned Jack again, and like an absolute legend, he came back for me. But in the time he took to drive those seven miles, the snow piled up some more. When we turned for home, the roads were impassable. We took stock, considered our options and adjourned to the pub. We went to the Cree-bridge Hotel. As the name suggests, the Creebridge sits on the edge of the River Cree. It started life as a country house, built in local granite, around 1760, for the Earl of Galloway. It has a well-stocked bar, good food and a big fireplace. Job done. We'd ride out the storm and wait. When we walked into the low-ceilinged bar, I bumped into Philip, my old neighbour from The Knock. Last time I had seen him I was knee-high to a grasshopper, and he was just one of the local dads. I had no idea how sociable he was then, or how much he liked a drink. So, there we were, holed up in the Creebridge, with not a lot to do. I probably don't have to tell you we got royally trolleyed.

Meanwhile, Jack's wife was at home, and she wasn't happy. He was out and she was jealous. She was determined to get to him. They had a Subaru 4X4, and she tried everything to get that car up the gravel road. She'd make some progress then slide backwards, but she kept going. She ricocheted the car off walls, as poor Ashley tried to talk her down. She phoned the Creebridge but was told that Jack wasn't there anymore. He was. We were just hidden in plain sight, drinking in a corner.

We stayed the night at Jack's sister's house, in town. The next morning Jack got a friend to drive us home in his Range Rover. Say what you like about their lack of reliability and their tendency to get stolen; that thing annihilated the snow. We passed cars abandoned at strange angles along the road. Back home in Creetown, the snow was six feet deep in places. The streets didn't get cleared for a week and a half. In the end, they had to dig them out. Meanwhile, you could walk from hedge to hedge, without touching the tarmac.

If the winter weather is going to wipe out your work and trap you somewhere for a few hours, at *least* make sure you have access to a fireplace and a stash of booze!

Chapter Five:

The April Fools' Club

In early April, 2025, for reasons best known to Mother Nature, we were getting an early shot of summer. Winter to summer in the blink of an eye, with just a hint of spring in between. For the past couple of years, we've had fantastically sunny spells in April, and this spin around the fiery ball has been no exception.

It was a Friday morning. A few months ago, we transitioned to working four days a week, Monday to Thursday; something I always dreamed of doing, even when it seemed like a long way off. Now it's a reality, and not just for me. Ashley, Craig, Graeme, and our latest recruit, Stephen, all work the same four-day pattern, which makes it all the more special. I like to tell people that teamwork makes the dream work, and a big part of that is rewarding the team. I've always wanted to create somewhere I'd like to work. For me, the four-day week only applies to trimming. Running everything else still takes up a bit more time, but it's nice having some breathing room, to focus on one less thing.

The boys were off school for Easter. The holiday fell late this year, so they got two weeks off, then four days back at school, followed by a long weekend. Not a bad deal. It was half past nine in the morning. The sun was shining on the front of the house. The dogs scurried round my feet, competing for attention. I was standing in my usual place in the kitchen, and I felt the need for coffee. In most of my Hoof GP videos, you'll see me drinking God-awful instant coffee. When you're on a farm, and a bit tight for espresso-making facilities, and Craig is your barista, you take what you can get. What I can get usually comes in a sachet, like sugar and chocolate with a side order of freeze-dried awfulness, but it gets me through. At weekends, and when I'm mooching about the house, the shiny coffee machine on my kitchen counter calls out to me.

We get to Glasgow as much as we can these days. It's a couple of hours drive away, through the Galloway Hills and up the West Coast. As far as shopping and entertainment go, Edinburgh might be the capital but Glasgow's where everything happens. There was a little café bar called Jelly Hill, on Hyndland Road, in the West End of the city. The traditional, glass frontage and rough-hewn wooden fixtures gave it a timeless feel, and you could sit outside surrounded by the red sandstone of the city and watch the world go by. I would go there every so often, when the well ran dry, and pick up as many kilogram bags of their coffee as they would sell me. I'm not even sure if selling the beans is something they normally did. I asked and they kept on selling them to me. I'd just found out that they'd closed, after twenty-two years, and it felt like the end of an era. I hope my supply lasts. Their coffee is so smooth, it goes down like hot chocolate.

I walked through the house, down the old concrete steps and out onto the terrace and the golden morning sun. The

Canada geese were honking out in the fields. It felt like they were here later than normal, though the effect was compounded by the unseasonal solar rays. Above the geese, I could hear the chatter of smaller birds, in the trees around our house. Ashley sat on the armchair opposite me, legs crossed, cradling her coffee in both hands and savouring the taste.

This is what life is about, surely? This is what all the hard work means, providing a place like this for my family, building something for the future. That's why we bought Carsegowan. It's a ramshackle row of barns, four miles from where we are sitting right now. I've only ever known it as a ruin. Thirty years ago, I'd pass it twice a day on the old double decker-bus we took to the Douglas Ewart High School and back.

People often ask me if I want the kids to follow in my footsteps. Do I want the boys to be hoof trimmers? I don't mind what they do, as long as it makes them happy and provides them with a life and a future. In some ways, I don't want them to go through a lot of the stuff I have. I love working on farms, but it's a hard life. It can be exhausting, and painful work, but it made me stronger and more resilient. It's the reason I have the life I do now.

On April mornings, the mist hangs low over the estuary, and drifts through the hollows in the fields, as we wind our way to work. The mountains in front of my house still had their dusting of snow, but the fields in the foreground were a more intense green. And even though recently, it seemed impossible, spring had arrived, and it was magical.

Scotland is famous for heather and thistles, but it's also known for spring daffodils; bright yellow flowers that grow in abundance up and down roadsides, on roundabouts, and in people's gardens. In the Highlands, they grow them commercially, on arable land. A few years ago, I was up north, and I noticed the fields were alive with them; row on row of glowing petals, in the shimmering sum. I said as much to a local farmer and his reply was sharp. 'Don't fucking start me,' he said.

My face probably gave away my confusion, and the farmer explained that when the daffodils bloom, it's too late to harvest them. This particular year, money had been tight, and they hadn't been able to afford the extra labour to bring in the bulbs, or the early sprouting flowers. In farming, timing is everything. If April is good and warm, like it has been this year, the cows can graze, and the milk will be good. The new lambs in the fields get off to a good start and, hopefully – although you can never tell with sheep as their mission in life seems to be to find creative ways to die – do well. In April, we can have driving rain or even snow. Plunging temperatures are hard on calves and lambs, and survival rates fall, along with the farmers' profits. For me, the sight of daffodils in the morning mist never gets old. Last year I spent a few days planting bulbs up and down the roadsides outside my own house, just so I have some daffs of my own. This is the first year they have appeared in force, and I love it. Just as well I'm not growing them commercially though.

April is an in-between time. You can sense new life pushing

up everywhere. The grass is thickening. The birds are vocal. The cows are spending more time in the fields, which, for me, means better feet! When I was a kid and farming with my family, this was lambing season. Dad took me out to help him, even when I was small enough to be a nuisance. Some of my earliest and best memories are of looking after pet lambs; the smell of wet wool and the bleating of the ewes in the distance; a constant soundtrack to my life. We would spend hours trying to pair them back up with wandering mothers. Some people I know call sheep 'woolly vermin', but I've always thought that was unfair. I love sheep. I prefer to think of them as 'field lice'.

Most mothers have twin lambs. That's the hope, but multiple births can make things complicated. My brothers and sisters and I would spend mornings on the farm quad bike; a Honda 'Big Red' 300. We would drive through the fields, grabbing as many lambs as we could. When we got a lamb, we'd spray paint a number on its side, then do the same to its mother, and any siblings. That way, we knew which was which, and where each lamb belonged. The fields sometimes felt like a living sudoku puzzle, with numbers everywhere, and five kids, trying to figure out which ones we'd already used.

Lambs were slippery little buggers, quite literally, when they'd just been born. You'd be running along, chasing one, and you'd finally make a leap for it, only to realise the lamb had switched direction. From there, it was a headlong dive and a face full of brown mud. In the farming community, there's a sense of pride in working long hours during lambing and calving season. We like to carp and groan about it, but if we're honest, those moans are a form of humblebragging.

Those lambing days were some of the happiest of my life, but they weren't without challenges. I remember driving down through a field one morning, with my sister Kirsty on the back of the quad. This particular field was called the Big Ben Boy. As with a lot of farms in the UK, all of our fields had names. Kaleb Cooper famously named every field on Clarkson's farm, putting the owner to shame. It felt like there were only two speeds on that quad: stop and flat-out. On this occasion, we definitely weren't stopped. We were speeding along, the wind whipping our hair and the bumbling growl of the Honda engine in our ears. And then, without warning, the nose of the bike shot upwards. The handlebars pulled hard left. I fought them, trying to steer right. I didn't know what was happening. I stamped down on the foot brake. The nose of the bike dived into the grass. The back end swung upwards, and one of the wheels overtook us.

I remember thinking, *Wait. What?* I very nearly lost control of the bike and my bowels at the same time. Somehow, a nut on the rear axle had snapped off, throwing the wheel away from the bike, and into the air. Mishaps, accidents and fuckups are common in farming. Farmers and people who work on farms need to be knacky. We learn to adapt and overcome, with what we have. You'd be amazed what I can fix with a piece of string or some bailer twine. There wasn't much fixing with that one though. I still remember the bike sitting, lopsided, dug into the dirt like a wounded animal. Kirsty and I were a bit winded, but luckily, we were all right.

I still deal with numbers on animals, nearly every day of the week, but nowadays I'm reading the numbers, rather than spray-painting them onto the sides of baby animals. We record everything we do to the cows' feet we look after. One day in 2019, I was working alone on a farm, when cow number six hundred and sixty-six emerged into the crush. I wasn't too bothered by the mark of the beast on her bum. I'm not superstitious and I knew I had probably put it there myself, back in my freeze-branding days. Maybe I should have been more concerned. I had a Wopa crush, then. It was my first good crush, but it wasn't as user friendly as my custom-made Appleton Steel or the KVK. Cow six-six-six was stubborn. None of my usual tricks were working, so I clambered in behind her, pushed my tired shoulder onto her dirty hips and heaved her forward, into the crush. Instead of moving to the front and locking into the head gate, as she should have, she threw her head out to the left side and through the bars. I pushed some more. She swung her head back to the right. Eventually, with a lot of coaxing from me, her head was in the right position. With one almighty push, I drove her forward and closed the headgate to lock her safely in place.

Drama over, I got on with trimming. I lifted her back right foot into the air. She kicked and thrashed and bellowed; angry at being confined by the steel of the crush. I picked up my grinder and got to work. I took care to go easy, then I reached for my knife and started to model out her back right foot.

Slowly, she calmed and settled as I worked away. I relaxed and focused on the sweeping motions of the knife. I held her inside claw in one hand and my razor-sharp blade in my other and pulled down through the horn. That's when six hundred and sixty-six lashed out, with a well-timed kick. The movement caught me off guard, and with all the force of an angry cow, the knife sliced through my nylon glove and what I was pretty sure was the end of my finger. I dropped the knife and grabbed my left hand. I didn't know if I'd cut myself. There was pain, but I wasn't sure about the extent of the damage. I relaxed my hand and felt a warm, almost burning, sensation as my glove filled with blood. With as much care as possible, I peeled off the orange remains of the glove. Blood spilled to the floor in a sickening splatter. I looked at my finger. It took me a moment to pick out the hairline cut. The blade was so sharp it had left a clean line. This was a problem.

I held everything as tightly as I could with my intact hand, trying to stem the flow of the blood, now a constant drip. My pinkie throbbed all the way up to my elbow. I looked back at the cow. Her foot was steady. She peered round the corner of the crush, back towards me, as though she was thinking 'mission accomplished'. Over the years, I've racked up knocks and bangs and scrapes in my life as a hoof trimmer, but this one stands out. We have a photograph to commemorate the occasion – a selfie of me and Ashley in the hospital. Ashley has a massive grin on her face. I'm in the background, wincing in pain. When the nurse cleaned my finger I could see ragged flesh, like globules of fat hanging out of the wound. She stitched everything back together as I tried not to squirm. She told me I'd nicked the bone. If six hundred and sixty-six had been more committed, she might well have had the whole finger.

The phrase 'if you didn't laugh you would cry' is never more apt than in the world of agriculture. You need a good sense of humour to get you through most days, and some farms need bigger laughs than others. One of those farms belongs to customer who has eleven or twelve other farms. He moves cows between his different locations, while buying in new stock from all over the country. That can mean you end up with a sort of mishmash of dairy cows, beef cows and everything in between. It's one of our favourite farms to work on, and one that has given me some of my best war stories. My brother-in-law Mark runs the place. We go there every four weeks, and they ferry in cows from their other farms, just for us; cows other trimmers have decided they would rather avoid or haven't had the time to attend to. We usually have a mixed bag: calm cows, stressed cows, cows who have never seen humans, and cows who have it in for me. They're cows that don't really belong together, or sometimes to any particular breed, a diverse selection of shapes, sizes and temperaments. To say some of them are wild would be a bit of an understatement.

A few years ago, I managed to squeeze enough money together to buy myself a new pick-up truck for work. A Nissan Navara. It was more than I could afford. I had to make a long trip to London and back, just to pick it up, but I was very proud. My new ride had big, chunky wheels and a manly bull bar; all the bells and whistles someone like me would want. It was a massive expense, but after a long run with a beat-up old work van I bought for a few hundred pounds, it felt like a big deal, like I'd finally arrived as a hoof trimmer. In my head, it made me look professional. I took pride in my work, and I wanted everyone to see that.

It was a reach for me, so it was important to keep my truck in good shape. The very first day I took it to a farm, I was careful. I parked up in a corner, well out of the way. As I drove away, the farmer – one of my favourite customers – held a gate open for me. Then, she accidentally let go of the gate, and I watched as it slammed into the side of my pride and joy. It was a small scratch, but this was only day one, and strike one, in what I didn't realise would be a very long list of battle scars. I certainly had no idea what was coming on day two.

We were trimming at the farm Mark manages, in a straw-bedded pen where we always set up. I moved my pick-up off to one side. It was a good eight or nine metres away from the crush, on the other side of two pens. There was no way an animal or a tractor could get anywhere near the pick-up. She was safe and sound, or so I thought. I had severely under-estimated my adversaries. We were trimming stirks: adolescent cows. Most of them weren't used to human interaction. That made them edgy. They're animals. They have no desire to hurt you, but their lack of experience means they're unsure about everything and everyone around them. And when you weigh half a ton, attack is often the best line of defence.

'Are you sure it's going to be safe there?' Craig said. He had only been working with me for a few months then. He too was very proud of the new wheels we'd just rolled up in.

'Of course,' I said.

All the cows were in the race. There was a good-sized gap between the race and the nearest pen. Beyond that pen was another pen, and beyond that, safely gated in, was my pick-up. We were about halfway through our day's work, when a small, dirty-grey, Charolais-cross bullock came prancing down the race. Behind our crush was the farm's own crush. We were using that as a holding pen. We would bring one cow down, put her into our crush and then have another standing directly behind us, ready to go.

This stirk was barely as big as a cow so he had more room than most, but when Craig moved him into the farm crush, he started thrashing from side to side. He ran backwards, then forwards, bashing into the gate behind me. I would never have admitted it at the time, but he was making me anxious. He was so wound up, and there was only a head gate between us. That's nothing too unusual. The standard tactic for the wannabe escapee is to try to turn back down the race and go back the way they have just come. If the race is wide enough, they'll poke the tip of their nose backwards and try to hurl their body weight around to face the other way. They are almost never able to do it, but this little guy was persistent. Try as he might, there was no turning.

'Nice try buddy,' Craig said.

I moved round to trim the front right foot of the cow I was

working on. That's when I heard the crash. I looked up. The bullock had charged the gate, bounced backwards and come to an abrupt halt on his knees. Stunned, he stood again and backed up. He rose like a stallion, on his hind legs. I watched, mouth open, waiting for him to tumble back to the concrete. That didn't happen. Instead, he executed a near perfect barrel roll. Right there in the race, he flipped, head over bum, and landed on his back. Then, fumbling and writhing and flopping on his spine, he found his feet and stood up again. Challenge complete. If Craig and I had been holding score cards up they would have been tens.

Should you ever find yourself in a situation like that, your best approach is to stand still and do nothing. I did that then, half terrified for what might come next, knowing he could easily catch a leg in a gate and break it. That would be the end of him. I wanted to help but my best option was to leave him and hope for the best. Back on his feet now, he was calmer. I turned to start trimming again, but he flailed his head and threw himself back up on his hind legs. This time, he pitched forward and up. I watched as he mounted the highest bar of the gate in the race. Craig and I stood there as he slid around on top of the bar, his bellows echoing round the open byre. And then, with a jerk and a flip of his head, he flopped forward, cleared the bar and landed in the first pen. If it hadn't been so frustrating, if I hadn't been so worried, I think I might have been impressed. It was quite a feat of athleticism for a bullock. Now that he found himself in his own larger pen, and with soft straw underfoot, he calmed down. Panic over, I went back to trimming.

Mark appeared and we told him what had happened. We agreed we would get the bullock at the end when he'd chilled out some more. I trimmed another four or five cows, letting the tension ease out of my muscles. I was midway through a trim when I heard another crash, an even louder bang, and Craig, shouting for all he was worth. Craig panicking isn't what you'd call unusual, but that doesn't mean it's unwarranted, and it always sets me off. I ran round the crush, following the noise. I pulled up short on the other side. The young stirk fish-tailed in front of me, gyrating his legs and his muscular hips. He had cleared the first pen and climbed into the second one. Now he was shimmying his way between the bars of the second pen AND MY PICK-UP.

I didn't want to accept what I was seeing. There he was, wedged between the pen and my beautiful machine. He wriggled and jiggled and kicked his way out of the small gap, as Craig and I looked on, helpless. Then, he righted himself and trotted away, with his nose in the air like nothing had happened. My first reaction was relief that he hadn't hurt himself. Then I looked at the rear end of the Navara, and the once pristine sheet metal. It now looked like a discarded crisp packet. It was gut-wrenching, like the loss you feel as a child when you break your favourite toy.

I looked at Mark. 'Aw, naw,' he said, as he tried, probably not very hard, to conceal a smile.

That would prove to be one of the smaller bashes the pick-up acquired in its life with me. In the end, I donated it to a worthy cause. I gave it to my good friend, Cammy Wilson, as a replacement for his own, even more banged-up grey

Nissan Navara, which at that point was held together with gaffer tape and hope. By the time I passed it on, it had seen things no vehicle should; like the time it got wedged on the side of my house, when I used it to move twenty tons of rotten rock around my garden, or the time I jumped out to help Craig unhook the crush and forgot the truck was still in gear and had to chase it and the crush through a shed. That race ended when the Navara got stuck in a gate and I landed on my face, in a pile of cow shit. I beached it three times: once launching my jet-ski, down the coast at Garlieston, once when I took the kids for a picnic at Kirkmaiden, where I grew up, and once when I took it to a beach further along the shore to see if it would get stuck with bigger tyres. It did, so I fitted even bigger tyres. I absolutely loved that truck, but I suppose if she were an ex-girlfriend she might have said I didn't treat her accordingly. I'd love to say she'll have an easier life in retirement but anyone who knows Cammy knows that's not the case. As I wrote this, he had just posted some CCTV footage of him jumping out of the pick-up, unhooking his trailer and driving away, right before the trailer rolled down the hill and into a tree.

By the end of April, the countryside is alive and blooming, and farming is busy. Everything points to the harvest season ahead. The Wigtownshire roads are sleepily alive. Slow-moving tractors haul machinery back and forth, trailing lines of traffic in their wake. This is our silly season. Farmers, farmworkers and contractors turn night into day, silaging grass for the year ahead, utterly at the mercy of the weather gods. I've always loved this time of year. It takes me back to when we silaged on our own farms. Now it could be done in hours. Back then, men and equipment would arrive, en masse, for three or four days. The dust rose into the air, as tractors flew up and down dirt roads, carting green grass from the fields to the silage pit, to be covered in a great black plastic sheet and weighed down with old tyres to ferment for the winter. I'd ogle the hardware. I'd take in the tractors, mowers and choppers and I'd dream of one day getting to drive something myself. I miss those days but love that I get to view them from afar, through my rosy filter.

I think of April whenever I smell freshly cut grass. That sweet smell lifts my mood every time. Time is tight, at that time of year. Up and down the country, farmers race to get their machinery ready. Lambs and their mums are moved off the silage fields, to let the grass grow in the April showers and the springtime sun. Days are longer and there is work to be done.

A face full of sawdust isn't my dream start to any day, but if I cast my memory back to April 2010 that's exactly what I got. I can still feel the sandy texture of the sawdust on my skin and in my mouth. The smell of cow dung filling my nostrils. It had all been going so well. There are more than a few tales of people telling others how to do their job, or what to watch out for when they *are* doing their job, and then tumbling into the same pitfall themselves. That's what happened to me, fifteen years ago, when I was still freeze branding.

People wonder if I get kicked by cows. The answer is a loud yes, but they're usually only glancing blows, something to wake me up, rather than knock me out. With freeze branding though, life was a little more precarious. Kicks are a regular 'perk' of that job. You're best making peace with it, preparing for the kicks and taking them on the chin – hopefully not literally – rather than going out of your way to avoid them. More often than not, if you do try to get away, the cow's hoof will connect with you at full extension, and you will be treated to the full force of the kick. If you stay in close, she can't get up quite as much speed. Yes, it'll hurt, but nowhere near as much as it does at full tilt. Think of me as a tired boxer, hugging my big, beefy opponent.

One cold, sunny morning, I was working on a farm ten minutes from home. Normally I'd have had help from a farmer or one of the employed workers, but on this farm, I would often turn up and realise I was alone. Rather than risk that again, I took Dave along for the ride. Like I said, Dave grew up farming, so he knows one end of a cow from the other. He also loves the gym. He's a former UK natural body building champion, so he's strong and up for a challenge; the ideal candidate, even if his type of farming mainly involves field lice.

We got set up for the day. The cows were in self-locking head yolks, all lined up at a feed fence. Each of them had an individual opening, through the bars on the edge of a long, concrete feed trough. When they put their heads through to eat, we could lock them in place. That kept them safe. It allowed me to work down the length of the shed, with the cows neatly packed in, side by side. It made the job faster than loading them into a crush individually, but it was a bit more precarious. The cows' heads might have been secure, but their bodies were unrestrained. They could move around, squirming from side to side. That was where Dave came in. When I was freeze branding, I used a special set of pliers we called humbugs. Imagine a large pair of nipper pliers, without the sharp ends. Instead, there are round balls, custom designed to fit inside a cow's nose. That sounds awful. I promise it isn't. It doesn't hurt the cow, but it does allow you to lead her by the nose. The cow will instinctively feel that if she pulls hard against the humbugs, it'll be uncomfortable, so she'll stand perfectly still and concentrate, rather than kicking the crap out of me. That's the theory, anyway.

I like to joke and play around with my friends, but safety around animals has to be a priority. We branded for a while, chatting away as we went. I've always been proud of what I do, and I was happy talking about work, explaining to Dave the hazards he could face if he was to come over to the wrong end of the cow, telling him stories of when it did go wrong. Dave was standing in the feed trough, in among the food the cows were trying to get at, humbugs in hand, grasping for the nostrils of each cow. It's tricky to catch them at first. There's a knack to it, but he got the hang of things quickly.

'They wriggle a lot. Do you not get kicked?' Dave said.

I told him about my up-close, slugger approach. I was worried he might get hurt, if he came in behind the cows and he didn't know the risks. I explained that in this situation, with all the cows lined up, side-by-side, it wasn't the cow I was working on I needed to worry about. It was her neighbours. My two eyes could only focus on one bovine bum at a time. The neighbours were further away; just far enough to build some momentum, if and when they struck. I told him I'd seen people laid out by the cows either side of the one they were branding, and that you needed to be careful. Obviously, I knew this because I was an expert, and I'd *never* get caught out like that.

My box of branding irons was three metres behind me, chilling on the rubber mattresses the cows slept on. I set everything up so I could work down the line. I prepared the cow in front of me. I shaved the hair from her bum, so the brand had full contact with the skin. I cleaned the area with methylated spirits, so that the skin and the follicles conducted the cold, and the brand would come out crisply. I talked to Dave as I pushed a set of cold, brass irons onto a freshly shaved behind. I held them in as tightly as I could, to get a clear brand. Dave held the humbugs in one hand and timed thirty seconds on a stopwatch in the other. Once we hit the magic number I removed the irons. I walked back to my ice chest, chilled to -82°C and replaced the numbers I'd just used.

Cycle complete, I pulled out the next two numbers and turned on my heels to find the right cow. As I turned, the cow next to the one I was looking for swung upwards, like the plastic mule in the boardgame Buckaroo. Her head was still in the yoke, held tight, and all that kinetic energy had to go somewhere. Her legs moved through the air, faster than I ever could. She was a small cow, and her feet were tight together when she struck me, right between the legs. It was a sickening combination. Two hooves, one groin, and a well-targeted kick, completed in a whipping motion.

I bent double, as my abdominal muscles contracted, and my stomach tried to climb up my throat. The branding irons clanged on the concrete. I stumbled, winded and coughing. I covered my face to shield it from more kicks. When none came, I looked up at Dave. He stared back, wide-eyed, like he didn't know whether to laugh, or scream, or start doing CPR.

'Are you okay?' Dave said.

'I'm ... fine ... I'm fine.'

'Take a moment. You need to sit down. I've never seen anyone as white as you are right now.'

I turned to sit on a rubber mattress, kneeled, then buckled, face first, into the sawdust. That was easily one of the worst kicks I've ever had the pleasure of receiving. I can still feel the grains of sand imprinted on my face. I can still smell the cow shit, right next to my nose.

It's not all about physical punishment. You get used to the kicks, but there's a mental brutality that takes a toll. They don't call it 'the grind' for nothing. The days are long, and the pace is not your own. Farmers are a busy breed, and they don't like taking breaks. You're really at their mercy. If you have two hundred cows, they may take it in shifts with one of their workers, but you're the one branding cow after cow, and taking kick after kick. It's a mental marathon, especially if you have ADHD, like I do. The monotony is hard. I don't mind *some* monotony. Some jobs have a rhythm that allows you to let your mind go. You can think about other things when you're cleaning the crush or doing repairs or sharpening knives. When you're timing brands and changing numbers, you need to concentrate. Looking back, some things seem funnier than others.

Cows are branded with numbers for herd management. When you think about it, branding is essentially a hardcore admin task. The numbers themselves correlate with the cow's passport. Yes, you read that right. Cows have passports; not for crossing borders or going on gap years, but for government records. It's an official record the cow takes with her between herds. It helps in the event of disease outbreaks, that kind of thing. All that means the numbers have to be bang on. Mistakes can cause big problems down the line. For me, the big frustration came when a farmer gave me the wrong number and I went ahead, taking them at their word. The number doesn't show up straight away. In fact, as soon as you take the iron off, it disappears. You can't tell what number you've put on just by looking, and the farmer might not find out for another eight weeks, when the hair starts to grow in white.

It was a lose–lose situation for me. If a farmer gave me the wrong number, I would brand it, and it disappeared for eight weeks. When it resurfaced it could cause havoc. If the number was wrong and that much time had elapsed, I wasn't around, and no one would remember the conversation. It was easy to assume it was my fault. So, I was always careful, listening for the number being called out.

'Number six,' the farmer or farmworker would shout.

'That's number six, right?' I'd say. I'd wait to hear a 'yes', and on the brand would go. I realise there's no one there to dispute this, so you'll just have to take my word when I say I very rarely got the numbers wrong. It did happen on occasion but 99.9% of the time my numbers were perfect. It's something I was – and still am – very proud of. It meant the farmers could rely on me.

Balvarran is a dairy farm, off a long scenic stretch of road in Aberdeenshire. They milk around four hundred cows on the farm. I visited once a year in the branding days, always in April. The young heifers were about to leave the shed for the open fields, and this was a last chance to get numbers on their bums, ready for calving time.

Dairy farmers replace 30% of their herds year on year, as calves mature into heifers. Angus, the man who owns Balvarran, generally needed me to brand a hundred cows every year. Working flat-out on a good farm setup, I could number a cow every three minutes, so I got through twenty cows an hour. When I say flat-out, I really mean it. Branding a hundred cows, pushing those numbers onto their backsides and holding them steady, and taking the inevitable kicks, is hard work. It was a tough five hours.

Just to complicate things, Angus didn't really trust his staff. He had a couple of guys working for him. There was Donnie. He was reliable but he was a bit of a lad, and I got the feeling he didn't have the boss's confidence. Then there was Chris. Chris wasn't too interested in his job. He was keener on smoking and avoiding anything that might involve problem-solving or long-term thinking. If it was day-to-day, fine, but anything bigger was a problem. Every April, I would turn up on the farm, and Donnie would offer to shout out the numbers for me, to save Angus the hassle and the time, and every year Angus would say no because he wanted Donnie free to push the cows up the race and let the work flow faster.

I know what you're thinking. What's wrong with a farmer caring enough to take on the responsibility of shouting the numbers himself; one who brings two employees to help keep things moving? Stop whingeing, Graeme. Surely that makes everything easier. It absolutely would have, if Angus hadn't had such an incredibly pronounced stutter.

As with anyone who suffers with a stutter or stammer will know, the more stressed you become, the worse things get. I'd be primed and ready, hovering over the freezing soup of numbers in my branding box. 'What number Angus?' I'd ask.

'Fif … fif … fif … fifty-six,' Angus would say.

'No problem. Fifty-six it is,' I'd say, hoping I was right. I'd brand number fifty-six, release her from the crush and then move on to preparing the next cow. 'What's the number, Angus?'

'A wu wu wu …Wu one one five …'

'Shit. Sorry Angus, what was that?'

'Aah … Wu … Wu … Wu … Wu one one five.'

'Aha, one, one, five,' I'd say to myself, and I'd apply the brand. I'd return the irons to their icy home, just as I'd hear Angus shout, 'Aaah … aye that was number fifteen … Yes?'

Oh shit.

I didn't confess to every number I got wrong on that farm.

I suspect it might have taken a while. Years later, I wonder just how inept Angus thought I was with numbers. I always tried to pre-empt the situation. At the start of each day, I'd politely suggest that Donnie and I could just work away and let Angus get on with something else. The reply was always the same. 'No, no. I like to make sure the numbers are all correct.'

Ah well, you can't win them all. Maybe May will bring simpler times.

Chapter Six:
Dying for a Shed

As I sat down to write this, Campbell was climbing the steel stairs in the new extension to our house. 'Quick. QUICK,' I told him. 'Run!' A huge roe deer was sprinting through the lush green grass in the field in front of the house. Something had spooked her, and she was galloping along the hedge. I tilted my coffee cup and offered her a 'good morning'. The sun's rays streamed through the big triangular window in front of us. Fluffy white clouds kissed the tops of the hills beyond the calm waters of the estuary and there was a warmth in the wind. Crows flew across the fields, foraging to build their nests or feed their young. A lady in a pink and white striped jumper strolled along the road in front of the house. It's never busy with traffic but people cycle past or wander with dogs throughout the day, and almost all of them shout 'hello!'.

If you're reading this, you'll know how much I love this country, and where we live most of all, but one thing we definitely lack is seasonal change. Our climate is tepid. We're on an island, warmed by the flow of the Gulf Stream. Draw a line along the same latitude, and you'll hit Germany, Denmark, Lithuania, Belarus, Russia, Alaska and Canada. All of those places have seasons. Our weather and our seasons meander along. The rain may be slightly warmer in the summer, but it's still rain. When autumn comes, it's a gradual process. Over six weeks, the leaves fall from the trees, carpeting the ground in golden-yellows, reds and browns. When winter fades to spring, there are false starts. We're hopeful, unpacking our shorts one minute and defrosting the car the next.

I don't think I'll ever understand when spring turns into summer here, but I don't think anyone else does either. It's the subject of speculation among the locals – an elusive point, like the source of the Nile. I suspect if we did hit on a definitive date, we'd miss talking about it. But sitting on my chair, watching the sunlight play on the water, it felt like summer was in the post.

These days, the sun lifts my mood, but years ago, when I first started trimming and I was still branding, I would have been thankful for the warmer, drier weather for practical reasons. Financially, things were a mess. I couldn't even afford a flat-pack garden shed to store my equipment in. I had a driveway, and every night I'd leave my work gear there, at the mercy of the elements. My mum says being skint forces you to make bad decisions. She's not wrong. I lived moment to moment back then. Always restless. Always on edge. The moisture in the air led to coatings of rain and frost and eventually rust on my tools. I had to replace things on a regular basis, which, in hindsight, makes a shed look cheap. After a year or two of this, I put aside £400 to buy a small garden shed. It sounds silly to say this now, but at the time I felt like I'd levelled up, like I did with the pick-up. Now, I was even more of a professional, with a workshop ... kind of.

I set everything up so I could take care of my things. In my five-by-eight-foot space, I laid out all my tools. I hammered six-inch nails into the mini rafters. At last, I had somewhere to dry out my workwear. I sat my electric grinding wheel on a table I fashioned from an old park bench. My shed was cluttered, but it was my clutter, and I knew where it all was. There was a system to the chaos.

After a year of this, slowly tweaking, evolving and teasing, I may have got a bit carried away. As you've probably worked out by now, that is often my downfall. I've never been keen on standing still. I like to move the game along. I like to tinker with equipment and streamline processes, and sometimes my enthusiasm gets the better of me. I can push a bit too hard. Sometimes that means breaking down barriers. Sometimes it involves a big jump in reverse and a liberal slice of humble pie, but whatever happens, I want to keep moving. I want to better myself, not just for material reasons – although I'd be lying if I said that wasn't a nice bonus – but because it keeps me interested. Coming up with ideas, testing them, fine-tuning; that's what gets me up in the morning. Those eureka moments, the tiny wins, are what it's all about for me.

In this particular moment of 'clarity', I had decided that my clothes weren't drying out properly overnight. How good would it be, I wondered, if I could heat this wooden shed that had already improved my life so much? I decided to make myself a wood burning stove.

I managed to pilfer a big, red Calor Gas cylinder from my stepfather's workshop – the kind of thing they use to store natural gas for space heaters. Davie was a prolific hoarder. In particular he liked to buy and hold on to tools. We once counted twenty measuring tapes in a drawer under his work bench, so he wasn't going to miss a gas cylinder. I made sure the cylinder was empty and cut a twelve-inch slit down one side – probably best not to try this at home, kids – then I welded on a pair of hinges. Once I'd added those, I made three more cuts to form a rectangular door. I was a happy man when I completed my final cut, and that door swung open. *It actually worked*, I thought.

I attached some feet and fashioned a chimney from an old gatepost. Perfect. You may be able to see the flaw in my plan, and, yes, it seems ridiculously obvious now, but at the time I was too taken with my shed heating solution to think about safety. The first couple of months went well. I'm sure I must have suffered a bit of carbon monoxide poisoning, sharpening my knives in my smoke-filled den, but I spent time in pubs in the early noughties, so I was used to that. With no major incidents, my confidence in the stove grew and – you guessed it – morphed into something more like complacency.

One Sunday, I set the fire and went inside for a tea break. I was padding round the kitchen in my socks waiting for the kettle to boil, when I glanced out of the window and saw flames licking round the top of the door, and smoke rising from the roof of my miniature cabin. It took a couple of seconds for my brain to register what my eyes were seeing. I stumbled to the door, tripping over my own feet as I tried to get my shoes back on. Outside, I tugged at the coiled garden hose trying to reach the blaze. I was lucky that time. The shed survived, even if it was a little more charred than I would have liked. I decided an electric heater might be a better idea in future.

Nowadays, the weather isn't such an issue. My tools don't freeze and thaw and rust, and nothing is covered in brown leaves in the colder months of the year, because I have the forty-by-eighty-foot shed; the one that appears in a lot of my videos. In summer 2021 we had just about finished renovating the house at Laigh Kirkland. It was the culmination of a dream we'd had for a long time, a permanent home for our family. But I was far from done yet. Having dealt with the inside, knocking down walls, putting up walls – not always the same ones – I turned my attention to the outside. We have an acre and a half to play with. I had plans for the lawn and the drive, and for fruit trees, but first I wanted a workshop. There were two sheds: a more general building, and a metal construction that sat low to the ground on the left-hand side as I drove in the gate. It had been built for rearing pheasants. It wasn't quite tall enough for storing and working on equipment, but it was in the prime location.

I engaged the services of Monkey, a local character and digger driver. Monkey pitched up with his digger and got to work, smashing the shed to pieces. He buried the biggest pieces of metal where I would soon be laying concrete. He would be a good man to have around if you ever needed to dispose of a body. We were left with a pile of wood from the innards of the shed. Not a problem. That could be burned, and it was a good excuse to indulge my passion for fire. Yes. My name's Graeme, and I am a pyromaniac. When I was a kid, my favourite job on the farm was burning whin bushes. They took a while to get going. The best technique was to push burning newspaper in as far as it would go and stand the hell back. The results were often incredible. I should probably point out that we didn't just do this for fun. It was a way to keep the whins in check, and it was good for the wildlife. Whin or gorse bushes get leggy as they grow, providing a lot less shelter underneath. I mean, it *was* fun too.

The wood from the shed was tinder dry as I piled it up to form a decent sized bonfire. I've never been big on patience. I like to get things done quickly, so I decided to give the tinder a helping hand and added a liberal dousing of diesel. I clicked the button on the lighter and got ready to vacate the area. This would be quite something. I stood and watched, in anticipation, as nothing happened.

Bugger it. I needed more kick. Then I remembered I had a jerrycan with some petrol in the shed that Monkey hadn't destroyed. That would get it going. It was a hot day, and I told myself to be careful of the petrol vapor. I poured on what I thought was a sensible amount of petrol, readied myself, and clicked the button on my lighter again. The spark ignited the petrol, the vapor, and me in the process. *I've just killed myself,* I thought, as I flew backwards, through the air.

I hit the ground with a thump. Flames licked up my right hand. I beat them out, then checked my body for more. There was nothing. I looked at my hand, swollen now and an angry shade of red. My face burned, like I'd been on a sunbed, but I was alive. Carefully, I got to my feet. I stumbled across the driveway to the back door of the house, wondering how I had gotten away with this.

I reached the step just as Ashley opened the door. She spotted me but she didn't properly see me. She pointed to some black rubbish bags at her feet. 'Can you put these in the bin, please?' she said. Then, she looked at me properly, and she started to laugh. 'You've got no eyebrows!'

Ashley closed the door. I stood there, winded, knowing she hadn't quite understood, and that none of this made sense to me either. I was probably in shock. I picked up the black plastic bags and loaded them into the bin.

Four years later, I still have the rough, scarred skin on the back of my hand. Happily, I have the shed too. In there, I can store everything I need, from pick-ups to power tools, to crushes, cars and the flags people have sent me from across the globe. I can work away, undisturbed, whenever I like. It's not something I take for granted, knowing where I've been. It's nice to be able to work without freezing or asphyxiating myself to death. That's not to say the shed is without its challenges. When we built it, we were security conscious. We fitted a big, steel roller-shutter door operated by remote control and a small black box that sits on the wall, just inside. The door rises like a giant sausage roll each morning and sits there most of the day before closing again at night. It's a fantastic idea in principle, but it has one fatal flaw; the same one we experience with other equipment we use; the squidgy organic matter that operates it, AKA me and The Craig.

The first incident occurred when I decided I knew the height of the pick-up intuitively. The door is electric, and you can pause it at any point in the cycle. I stopped it at seven or eight feet, knowing I had plenty of clearance. By not raising it to the top, I decided, I would save myself a few seconds. Another moment of glorious ingenuity. We were about to set off on the six-mile journey to a farm along the road. We drove out under the ajar door, and I was pleased when the pick-up cleared it easily. My smug grin lasted maybe a second and a half. As I turned right to exit the shed, I heard the almighty crash of metal on metal. The pick-up sprang backwards. Craig and I whiplashed forward, then back into our seats. That's when I realised I had judged the height of the pick-up perfectly, but I'd forgotten that the crush was a good two feet higher.

With a sagging head, I jumped out of the pick-up to survey the damage. The top of the door looked okay. The bottom section bowed out in a wide curve, like a metal banana. I had surpassed myself. I'd even managed to rip it clear of its rails. It stood proud, in mid-air, taunting me. I was annoyed with myself. What a stupid mistake to make. I looked at Craig. Craig looked at me. There wasn't much we could do here, so we went to work, to see what else the day had to throw at us.

It took me two weeks, but I managed to get the company to send someone out on the five-hour round trip to fix my mess. Things got back to normal. We had a door that opened and closed – which now felt like a luxury – but not for long.

This time it wasn't my fault. Craig, if you're reading this, you know deep down that I'm right. Just accept it.

We were on our way home after a particularly long day. We had been down near Drummore, on the southern tip of Scotland. It's a wild place that feels a long way from anywhere. Stranraer, the nearest town, is twenty miles north. A very good friend of mine – I won't name him – calls it Craggy Island, after the fictional setting in the nineties sitcom, *Father Ted*. That fits pretty well, I'd say. You really feel like you're out of it, down there. We had trimmed around a hundred and twenty stubborn Montbéliarde cows. That's a big day for us. We normally get through forty cows in a day. The herd were in fantastic shape, so the trimming was easy. *Getting* to the trimming, not so much. Montbéliardes are dual-purpose cows, bred for milk and beef. They're stronger than most dairy cows, and they have the sort of nature you'd normally find in a curmudgeonly old bull. Getting every one of them into the crush was a battle.

By the time we made it home, we were both knackered. It had been a hot, sweaty day, under a concrete roof, broiled by the sun, then an hour's drive home. I pulled into the yard and opened the roller door with the remote control, then reversed back into the shed. I stopped just as the pick-up's bonnet – or hood if you're in the USA – reached the door. I remembered to lift the door all the way this time, wanting to avoid further calamity. When we get back to the workshop at the end of each day, we spend time going through the equipment and what we need for the following morning. We potter around, sharpening knives, cleaning and tidying everything so we're good to go.

On the day in question, I couldn't reverse all the way back into the shed because one of the kids' buggies was parked behind me. The crush was in the shed, and the pick-up was most of the way in, with only its nose sticking out. I left it that way while we did all our end-of-day chores. Then, when we had everything sorted, I moved the dirt buggy out of the way, ready to pull the pick-up safely into its resting place for the night.

Craig was still busy, sharpening knives, when I climbed back into the driver's seat to begin my manoeuvre. Reversing a trailer is an acquired skill. Sometimes you need to shunt forward to realign the towing vehicle, to place the trailer exactly where you need it. I slid the gear lever into drive, and pulled forward, out of the shed. I turned the wheel as I went, to get the angle just right. I had been reversing trailers for years, so this wasn't something I had to put much thought into. Or so I thought. I was in a hurry to get done for the day and get everything locked up for the night. As the wheels turned in, I put my foot down. Then, I heard the now familiar grinding screech of rapidly contorting metal. I careered forward as crush met door, once again. I wasn't going anywhere, so I wasn't wearing a seatbelt. My face hit the windscreen, my foot hit the brake, and the momentum of my body mashed my knee into the steering column.

This was more of a shock. It didn't make sense. What had I done? I knew I had opened the door fully, so how could it have caught the crush? When I got out though, there it was, that same mangled mess. The door had burst free from its runners once more. But how had it happened? Had the door closed itself somehow? No. It had not. Craig had decided a fully open door would let too much wind into the workshop, and so, in order to keep everything cosy, he'd decided he should lower the door to just above pick-up height, but well below the top of the crush. That would have allowed me to get in and reverse straight back, if I was *going* straight back. Craig didn't know I wasn't, and I didn't know about his bright idea. I didn't check how high the door was because I didn't think I had to. Craig still blames me for not checking and I still blame him for what was a crackpot idea in the first place.

I reluctantly called the door company. I should have made Craig do it. I tried to make light of the five-hour journey they were going to have to do, again, but there was a long silence on the other end of the line. I filled it with the sound of nervous laughter. These things happen – especially when Craig's around – and you can't get annoyed about lumps of metal. YouTube has been good to me and money isn't the problem it was in the beginning. I don't see the point in getting angry about stuff you can fix.

When it comes to cows, I'm a lot more easily rattled. I get upset, sometimes angry, when anything goes wrong with the animals we see. Yes, they're huge – often weighing over a ton – so you'd be forgiven for thinking they were indestructible. But you'd be surprised how easily they can hurt themselves when we're working with them. Safety is king. We don't want anything to go wrong. It would be easy for me to sit here and tell you it's just because I care, but, selfish as it might sound, I also don't want to be the one phoning a farmer and telling them we have killed one of their animals. All difficult calls take their toll, whether you're delivering bad news or chasing payment, but delivering avoidably bad news that costs someone money, and hits them in their livelihood, brings up a whole other level of anxiety on both sides.

Seeing animals suffer is the worst thing that can happen in my job, especially if it's the result of them coming into contact with us. When things look like they could go wrong for a cow, my heart and mind race. It's probably the only time you'll ever see me flustered. I'm desperate to avoid the cow hurting herself, and to keep us all out of harm's way. I'll be the first to get stuck in when these situations arise. Craig and Cameraman Graeme – and anyone helping us – are my responsibility. Don't get me wrong, I never want to get hurt, but I'd rather that than see anyone else come to harm doing their job. Actually, what am I saying? I've had quite enough injuries!

There's a gorgeous village on the end of the Argyll Peninsula, called Machrihanish. The village is home to some of my oldest customers from the freeze branding days. If you're a keen golfer, you may have heard of Machrihanish Golf Club. It's a links course, designed by Old Tom Morris, the grand old man of golf. Jack Nicklaus declared the first hole at Machrihanish the greatest opening hole in the world.

It's a place of pilgrimage for a lot of golfers, though the area itself is arguably more famous for the Wings song, 'Mull of Kintyre', Paul McCartney's romantic tribute to the area's beauty. I can see what he was thinking when he wrote it, but romance is the last thing that comes to mind from one particular trip down there.

We were freeze branding some highly strung yearling heifers – around forty or so – in a long race formed from shuttered concrete on one side and galvanised steel bars on the other. I love working with long races. It means we can line the cows up, behind me, ready to go. It makes the day run more smoothly and efficiently, saving everyone time. The sun was sitting high in a clear, blue sky and the world seemed in its rightful order. There were four of us working that day, and we were getting through the numbers at a decent pace. As the cows weren't fully grown, I had a bit more wriggle room than I would have liked.

We'd chase one cow up into the head gate, and momentum would click the steel bars in place in front of her shoulders. I'd slide in through a small gap in the railings, leaving me perfectly positioned behind the cow. Looking at it from a cow's point of view, nothing good happens when you have your head locked in a steel gate. Unless you like injections, or having your feet trimmed, or being branded, it's a tough shift. These girls were riled up even before I pushed my ice-cold branding irons into their flesh. Cows bellow when they're in any sort of distress. It's a call to the rest of the herd, which makes sense in the wild. In this situation though, it isn't ideal. The noise only heightened the tension in the race behind us.

I would usually brand cows on the right bum cheek. With cattle so small, I had to crouch down on one knee, ducking low into the right-hand side of the race, directly behind each cow. It was a precarious position, squatting down at the wrong end of the animal, smack bang in the heart of the kick zone. I tensed my upper body, focused on keeping the irons in place, to get a good clear number. Instinct screamed 'get out of there', as the kicks flew past the side of my face, and into my thighs. I only had eyes for the cow in front of me. I needed to get the job done and get out of there in good shape. Jim, one of the farmhands, looked after the cows in the race, moving them forward. His most important job was slotting a fence post in behind me, when each cow moved forward. That stopped them coming too far and crushing me in the process. Always a bonus.

The race emptied and I was on the last cow, so Jim went to get some more. Doubtless he was more concerned with keeping things moving, but he didn't replace the fence post. It was a momentary lapse in concentration. Meanwhile, I was in the zone, fully engrossed with my current 'client'. I didn't miss the post, because I wasn't looking.

I heard the rhythmic pitter-patter of hooves, splashing their way down the mucky race behind me. I realised something was off when the pace didn't change. The cows behind me weren't slowing down. I spun my head around in time to see a pair of front hooves rise up towards me. Manure splashed my cheeks and nose. I felt the warmth of hairy front legs cross my face. I dropped my branding irons and threw myself on the ground. I crouched low in the stinking slurry, arms over my face. My head bounced off the concrete wall, and my eyes caught sight of the cow's back feet as she ran right over the top of me. I clambered up from my position and looked up, to see her launch herself onto the back of her herd-mate. As I got to my feet, I was thankful it hadn't been worse, but more concerned by the scene in front of me. The second cow was still on top of the first. The bottom cow was unstable under the weight. Her knees gave way, and she buckled, then collapsed in a heap. The cow on top craned her neck round to look at me. *What the hell happened there!?,* her long, black and white face seemed to say.

If I'm not creating uncomfortable, frustrating situations for myself, it sometimes feels like someone or something else, or the universe, will create them for me. I suspect that's a recurring theme in all of our lives. Over time it has led me to understand that no matter how bad a situation seems, there's always a solution, even if it's not the one you were expecting. Life moves on and, hopefully, tomorrow will be better.

In this case, I'm not entirely sure how the cows untangled themselves. I calmly replaced my irons in their icebox and told the farmer it was time for lunch, and I would see him in half an hour. Sometimes, you need to know when to walk away. The headgate was unusual and the cow on the bottom would have to back out of it so it could be opened. The solution would need to be unorthodox. For what it's worth, I *do* remember it involved two telescopic handlers, a sledgehammer and a lot of shouting.

Perhaps not simpler times after all, but hey, at least the summer weather is coming, and the ground will be a bit drier under my feet.

Chapter Seven:
Chinese Calamari

China should have been straightforward; a literal long haul, yes, but uncomplicated. Grit your teeth, board in Glasgow for the first leg, quaff some complimentary bubbly to tide you over until a quick changeover in Dubai, then on to China. Nothing more complicated than that. It was June 2025, and as usual the brown stuff had hit the fan, spraying sideways and covering me in the process.

Davie McGarvie was my stepfather, my mother's husband; he was an arsehole at times, but one with a heart, and yeah, I realise that's a troubling mix of metaphors. Davie's the reason I started trimming cows' feet all those years ago and a very big part of the reason I still do today. He taught me many, many things. For example, if you bop a bull over the head with a rubber mallet because he's misbehaving in your crush, do not be surprised if that mallet bounces back and breaks your nose. Also, if you wish to donate to the community of the small west-coast village of Ballantrae, you could do worse than buying them half a dozen very large, very expensive pot planters for the roadside, especially if you've been distracted while pulling your crush along their main street, and accidentally annihilated the originals that have sat there, untroubled, for years.

These were valuable lessons, but there was an unfortunate recurring theme. Most of what Davie taught me was accidental. I run my current business in the way I do because I was able to learn from his mistakes rather than his successes. In some ways, I'm standing on the shoulders of a giant, even if he was only about five foot six.

His work ethic was incredible, and something I've tried to emulate. It's beyond my reach, if I'm honest. Family is my priority, now, and for that reason, I can never hope to achieve what he did in a week, but I don't want to or need to do that. Davie fell victim to the idea that hard work was the be-all and end-all. He thought that almost killing himself with graft would somehow equate to riches down the line. On one notable occasion he said he'd like to die working. He subscribed to the notion that if you just get your head down and work, if you ignore everything else, you will be rewarded in the end. It's a seductive thought, but in my experience, it doesn't really work like that. You need to focus on the long view.

Davie often fell into the trap of thinking of his customers as friends, rather than clients. He went out of his way to keep people happy. He did jobs at cut-price rates and travelled across Scotland to farms miles from any other job he had going on. He was never able to bill for half the hours he worked, driving out of his way, or doing extra jobs as a favour. In the end, he spent more than he made. It's not always about how hard you work, though you do have to work hard. Sometimes, you need to be smart about it. You need to find ways of leveraging all that slog and making every last minute count. If you succeed in doing this, tell me how!

For me, the trick to making the most of the hours I work is through building something that grows a life beyond you. Whether it's a brand or a product or a business, I really believe that's the key to being able to provide my family with everything they need and to be around, to be present as a husband and father, and to leave something behind. That's how I found myself in seat K9 on a colossal Emirates A380 aircraft.

K9 was a bit fancy, bound in sumptuous leather and embellished with gold and walnut. I was headed for a town called Ningbo. Ningbo is a 'town' with *seventeen million* people, around three hours from Shanghai. We sell our own hoof glue, branded as Hoof Grip Pro – see what we did there? – and this trip was a journey into the unknown for us. We were there to promote Hoof Grip Pro to the emerging Chinese dairy industry.

Time is always tight for me, and four days were my limit. I say 'were'. I went to Glasgow Airport three hours early: the two hours you need to be there beforehand, plus an insurance hour. My ADHD has left me with a fear of being late. I waited five hours beyond the time the flight was due to leave. I hung around in the sickly fluorescent flight of the airport for eight hours in total before they finally cancelled it. Then, the airline rearranged all the other passengers' flights and booked everyone into a hotel. Everyone except me. *Fuck it,* I thought. *This is a bust. I'm going home.*

I climbed into the Raptor and drove for an hour, down the west coast, then turned inland, along the snaking, single-track roads of the South Ayrshire hills.

Meanwhile, Dwayne kept messaging me. He was in Shanghai and wanted me to turn around. 'Please, please come,' he said. Dwayne is not one to do pleading. He's the CEO of Hooftite, the company we partnered with to make Hoof Grip Pro. Dwayne is an imposing figure in every way; the sort of shape you'd sketch if someone asked you to draw a 'big bloke' after a couple of pints. He's six foot six if he's an inch and built like a wardrobe that just swallowed another wardrobe. He's frighteningly intelligent, questions everything with the enthusiasm of a toddler discovering jelly for the first time, and I can't help but grin like an eejit[3] whenever I'm talking to him. Saying that, much as I like Dwayne, my decision also had quite a lot to do with a phone call from a kind lady called Vanessa, at Emirates. Vanessa turned out to be a YouTube subscriber, and after some pleading on my side, she upgraded me to first class.

I was assuming more would yet go wrong on this trip, but right then, in my big, comfy seat, in front of a screen, I was happy with the perks-to-problems ratio. I wondered how Davie would have felt in my shoes. I suspect, if he'd been there, there would have been swearwords. He couldn't do enough for people he knew. He would have helped anyone out. His catchphrase was 'nae bother'. Nothing was a problem. But his temper was quick, and the little things often escalated.

---

3. 'Eejit' is a Scots word for idiot.

Someone – usually me – would inevitably break something, and the fall-out would be worse than the actual damage.

He had an old, red Nissan Terrano, a small Japanese 4x4 that had seen better days. Even the speedo didn't work. We had to use the GPS speed reading on the satnav to avoid the long arm of the law. A favourite game we invented involved opening the sunroof and driving along flooded roads at speed. If you timed it right, the unsuspecting rear passengers got hit with a wave of muddy water. I took the Terrano out to run some errands one day and came back with a shattered windscreen. By the time I got to Davie's workshop, it was a floppy mess, held together by the laminate surrounding the glass. I'd told Davie that as I'd been driving along, an oncoming vehicle had thrown up a stone, smashing the glass. Davie's face turned an unusual shade of purple.

'How can you be angry at me for something that wasn't my fault?!' I said.

'It's NEVER your fault,' Davie barked back.

That was over ten years ago now, and I'm still frustrated that he didn't believe me. Okay. Technically speaking, I was bending the truth. Quite a bit. I'm just glad he never found out that I'd had a twenty-minute dual with a huge, buzzing, bluebottle, as I drove along the local back roads. I lashed out, trying to swat it. I missed dozens of times, but when it paused for just a second too long, BANG! I slammed my fist as hard and fast as I could, straight through the glass. In the white-hot heat of battle, I had failed to consider the consequences of punching the fly *and* the middle of the windscreen. Still, Davie thought it was a stone and blamed me. Unlike the bluebottle though, he is very much missed.

I'd say June is my favourite time of the year in The Shire. The sun brings real heat and the world around me is alive. The air is a heady mixture of sweet scents and my ears ring with the meh-ing of lambs playing in the steep banks in the field behind my house. The birds sing from dawn until dusk, and my kids are in the garden more often than not, which, as any parent knows, is the dream.

Scotland's hills, moors and glens were shaped in turn by fire and ice, as colliding continents and volcanic activity drove up the hills, then receding glaciers washed away the valleys. The weather that sweeps across our land, the rain, and a climate warmed by the Gulf Stream give the fields their intense, green hue. Whenever I catch myself wishing for more sunshine, I remind myself that all this beauty is only possible because of the rain. That said, like most places, whatever thousands of Instagram reels might tell you, Scotland is at her best on a sunny June morning.

There's no place I'd rather be than at home in my garden. I mentioned this before, but every summer we have a family barbecue. It's a get-together of aunts and uncles, brothers, sisters, nieces and nephews. It's generally a drunken fun-filled mess, and that's how it should be. I am a child at heart. I just can't help myself. I fill the garden with inflatables – for the kids, and maybe just a few of the grown-ups – and we come up with some home-spun entertainment. Last year turned out a little unfortunate for my mother-in-law.

We'd had a good run of hot weather, so I got the kids' pool out. It sat there on the patio at the front of our house. I can't even remember what led to it. Predictably enough, Ashley's mum, Mary, and I were winding each other up; all harmless banter, normally. Eventually, probably because I was losing the argument, I threw a protesting Mary over my right shoulder and made a run for it. I loped through the collection of laughing and cheering in-laws and launched my mother-in-law into the ice-cold pool. I was quite pleased with my aim, and the reaction my stunt got me. Everyone was killing themselves laughing. It took me a few seconds to realise that they weren't laughing at me though. They were laughing at Mary's state of undress as I carried her over my shoulder. Let's just say I'd upended her. Her modesty had been compromised, and some of her womanly assets – specifically, the left one – made an unscheduled appearance. Mary had the last laugh though. She's in pretty good shape and picked up a few compliments across the rest of the afternoon. Sorry again, Mary.

My flight was supposed to take nineteen hours, with a two-hour layover in Dubai. The new flight meant my two-hour layover in Dubai turned into a nine-hour wait. I scanned the pop-up minibar to the right of my seat and realised it was all soft drinks, so I ordered a cheeky glass of Dom Pérignon. Needs must. It was the middle of the night when we landed, and they decanted us into the terminal in Dubai. If it had been daytime, I would have headed into Dubai proper, looking for some kind of entertainment, or just to wander round the shops, in the heat. As it was, I spent those extra hours in a feverish dream, thrashing around in an uncomfortable chair, waking again and again to the sound of a loud woman who just wouldn't stop talking.

Back on the flight, for the Dubai to Shanghai leg, I went to the toilet, and when I emerged, there was a woman lying prone across the aisle in front of me. I wasn't sure if she had passed out, but she was lying there in a pile of puke. One flight attendant was on her knees, helping the woman, while another hovered in the background, passing tissues and things back and forth. I didn't think this woman was drunk, but I'm not sure what I was basing that on. She lay splayed across the aisle, and she was a mess. It was hard to imagine a worse outcome for her, bar the plane actually falling out of the sky. The woman was wearing socks that said 'No Bad Days'. I had to fight the urge to take my phone out of my pocket and photograph the whole sorry scene, just so I knew this was real.

And then we were landing in Shanghai, and it was incredible. KVK had sent me a driver. He held up a card with 'Graeme' written on it.

'Are you for me?' I asked him.

The man held his hands up and said, 'Graeme', but he was wearing a KVK branded hat, so I guess that narrowed it down a bit.

'Yeah. You'll be for me,' I said. 'Is your car over there?' I pointed in the direction I thought it might be.

'This way,' the man said, and he made a grab for my bag.

'I can carry it,' I said, but the driver wrestled it out of my hand and started walking.

'How many hours to the hotel?' I asked, hopefully.

The driver offered me a blank look and that's when it hit me; he didn't really speak English, just enough to collect me, which was still a lot better than my Mandarin.

'My name is Graeme,' I said, jabbing a thumb at myself. 'What's your name?' I pointed to the driver.

This too was lost in translation, but he turned and started walking with my bag, so I followed him outside, into a pitch-black night and towards the car. I'm a petrolhead. I love cars, but I had no idea what was sitting in front of me in the pick-up zone. I could tell it was electric, which – to me – put it in the same category as a blender or an air-fryer. I climbed into a back seat shaped like an armchair and made myself comfortable, not knowing quite how long I was going to be there. The interior felt high-end, like a Merc, or maybe a supermarket's-own-brand version of a Merc. A hand emerged from between the front seats, bearing a bottle of Heineken. 'Beer?' the driver said. Now we were talking the same language.

'Yeah, I'll have a beer.'

It was dark outside. I couldn't see any of Shanghai. I typed 'Sheraton Hotel, Ningbo' into Google Maps. Google told me we were two hours away. I glugged my way through more beer and stared out into the night. I checked my phone intermittently until it told me we were ten minutes away.

'WC?' the driver said.

'No, no, no,' I said. All that lager did mean I needed to go. *But why does he want to stop when we're so close to the bloody hotel?* I wondered. We pulled up and he got out for a cigarette. I told myself that made sense, sort of. I've never smoked so I don't really understand the urge. I'd rather just wait until I was in my hotel. It had been a long journey, and I was tired. I just wanted to get there.

We were twenty minutes down the road when I checked my phone again. We were getting further away from the hotel. That's when I realised there are three Sheraton hotels in Ningbo. And so, I sat for the next hour and a half, legs crossed, regretting my choices. It was 1 a.m. when we arrived, but as far as my hosts were concerned, the night was young. Palle and Dwayne had decided they would make me feel at home by meeting me in a Scottish bar. Palle is the CEO of KVK, the company that makes my bright green cattle crush. A neatly dressed Danish man who always seems to be smoking, he's very deliberate. He takes his time to think, and to say what he means. He might be my polar opposite. The pair had told the owner, a lady called Jenny, that I was coming and that I'm from Scotland, so she had decided to keep the place open, just for us. So, there we were, Jenny, two members of staff, me, Dwayne and Palle, two hours into a lock-in.

Someone handed me a mule when I arrived. I was already a few beers deep but I decided this might be the best thing I'd ever tasted. It really hit the spot, lulling me into a state of amiable relaxation. I'm not normally a whisky drinker but the tiredness and the disorientation of being a stranger in a strange land had interrupted normal service, and I got stuck in. They had three hundred or so whiskies behind a bar, backlit in red. Jenny knew about our local Bladnoch single malt, and she seemed to know the area. She was such a good host that when we eventually staggered out of her bar, it was 7 a.m. We were all steaming drunk on whisky, on a Tuesday morning, and I felt quite patriotic.

Back at the hotel I managed two and a half hours sleep. When my alarm went off it felt like my head had just hit the pillow. I was shattered and twitchy, well into my sleep over-draft, but this wasn't a holiday, so we were up and into Ningbo. In the daylight I was struck by the height of the buildings and how clean everything was.

My first impression of KVK's Chinese team was that they are all nuts. We arrived at their Ningbo factory, and they had rolled out a phosphorescent green carpet. They had confetti canons and steel cut-outs of women in bikinis. Women in bikinis with my face! The scene was surreal, but *everything* was right now. In the office, we were down to business quickly. Our day was jammed with meetings about social media strategy, how to push hoof-trimming forward in China, how hoof glue works, and a tour of the KVK factory. Our Chinese counterparts were enthusiastic, engaged, and keen to learn anything they could. They give us a presentation they had been rehearsing for a month.

Here's the thing about China; everything I thought I knew was wrong. Or it seemed to be. I thought it would be polluted. I thought it would be overpopulated. I thought in-dustry ran on slave labour. I thought it would really *feel* like a communist country, or the grey and brown sludgy version of a communist country I had in my head. One with propaganda posters and queues. I thought I would feel intimidated by everyone and everything. I was very, very wrong.

For starters, they've banned all two-stroke engines. One day, they just banned the whole lot. So now, instead of the ear-splitting miniature petrol engines you naturally expect to buzz around you in Asian cities, like metal wasps, you have mopeds that don't buzz at all because they're electric. There are thousands of people around you on scooters, and they're lethally quiet. I found myself walking down the pavement, with Palle grabbing me and pulling me out of the path of an oncoming scooter. The traffic may be cleaner, but the scooters still go anywhere they want. The giant zebra crossings are a white-knuckle ride. Ningbo is a substantial city, but it felt calm and relaxed, disconcertingly quiet. Sixty or seventy percent of the cars are electric too.

There are over 1.5 billion people in China. The traditional Chinese diet isn't high in dairy produce, but that is beginning to change. When we think of dairy, we think of milk, but the bigger part of it isn't liquid at all, it's powder. Milk powder is used in a long list of products; baby formula milk is the obvious one, but pasta, cereals, ready meals, chocolate, bread and baked goods all use it. In the past, China imported a lot of milk powder, but now, having taken a decision to rely on the rest of the world less, they are all-in on dairy.

In many ways the Chinese are years ahead of us, but in cattle husbandry, and in the care, cultivation and breeding of cows, they are keen to learn. In my world, for example, they do trim cows' feet, but more from a maintenance point of view. They don't do it therapeutically, yet. If a cow has a problem and goes lame, farmers don't have the skill or experience to know what to do, and that can mean the cow dies, needlessly. For someone who loves animals as much as I do, that seems an unnecessarily harsh reality. From a commercial point of view, it's also highly unproductive. It doesn't make sense, when you can fix cows or stop them going lame in the first place. There are lots of things we can do that the Chinese dairy industry isn't familiar with yet. Blocking is a good example. If a cow has really bad bruising, she might not be lame, but it's an early warning. Lameness is on the way. If I trim a cow's foot well, she's still got bruising, and she'll keep on bruising, so I need to get weight off that bruised claw. If I hurt my foot, I'd use crutches. With cows, we use a block to add some height to the healthy claw and keep the bruised claw clear of the ground. That's not standard practice in China ... yet.

Don't get me wrong. It was very cool to be there and to help Chinese farmers in any way I could, but this wasn't altruism. I wasn't there out of the goodness of my heart. China is a massive market with big opportunities, and if they weren't into blocking cows' feet yet, that meant no one was selling glue. The team gave us an introduction to Chinese social media. In the West, we talk about China having a social media ban, but that's not quite the full story. They just have very different platforms. They can't access Facebook, but WeChat, Weibo and Douyin – AKA TikTok – are all massive. WeChat feels like as much of an extension of your personality and your being as anything ever could be. A billion people use it for everything, including payments.

The KVK team explained all of this to me in the meeting, and we bounced ideas back and forth. Having an existing catalogue of material would help me. And, it turns out I've already had a kind-of head start. I might not have signed up to any social media apps personally, but my videos have been pirated and uploaded everywhere. I have become Chinese-social-media-famous without doing, or even *knowing*, a thing about it.

We followed up the first meeting with a tour of the KVK factory. Rows of green crushes stood in line, like soldiers, ready to join the fight. It was a satisfying sight. KVK is a Danish company, and they produce a lot of their crushes in Denmark, but ten years ago, in a moment of foresight, they set up a factory in Ningbo to cater to the emerging Asian market. They were big on detail. So much so, the factory here is set out in exactly the same way as the one in Denmark. I realised they had taken it pretty far when they told me they had renamed each of the Chinese employees to match their Danish counterparts. There was an Andrew and a Johnnie, neither of which sounded very Danish to me. KVK don't do things by halves. It's that attitude and attention to detail I love. That's why I use their equipment.

The big reason I was there was to meet people; to put faces to names I'd be hearing more from later on. This was an introduction. We held an impromptu press conference, and it wasn't what I imagined. I had been worried that I wasn't bringing enough to the table. KVK and Hooftite are very friendly as companies. KVK is helping the Chinese farming industry move forward with hoof trimming, because that helps them sell more crushes. In order to do that, the industry needs glue and blocks. So, Palle got in touch with Dwayne and his business partner, Brent, and said, 'do you want to sell your glue?' At that point, Dwayne and Brent got in touch with me and said it would make more sense if we sold Hoof Grip Pro out there. This was a big, combined push, with KVK, Hooftite and The Hoof GP all going in together.

What I hadn't realised, until now, was just how many people in China already watch my pirated videos. I counted five regional news outlets there to interview us. Somehow, despite the beers and the mules and the whiskies, it all went well. I managed not to say anything stupid or offend anyone. It was only when the weight began to lift that I realised just how much fear I had brought into the room with me. That, and a big dose of imposter syndrome. I had flown to the other side of the world and all I had up my sleeve was the hope that some people might have seen my videos. But no, it turned out *a lot* of people had seen them. In the end, we were on twenty-seven news channels, and our interview hit the two biggest channels' viewing targets for the month.

Dinner was in a traditional Chinese restaurant. To my eyes, it looked like a very posh traditional Chinese restaurant, with wood-panelled walls and a door leading to a balcony. In Ningbo, it seems, you can get anything you want: McDonalds,

Burger King, all that stuff. American, English and French restaurants, anything you can think of. But I was happy to see the Chinese team at KVK were proud of their roots and they wanted to keep it local.

There were the three of us: Palle, Dwayne and me, plus eight or nine people from KVK. We were in a private dining room around a circular table. The piece in the centre turned, like a lazy Susan, but bigger, and packed with different dishes. I knew I needed to show willing. I needed to embrace the local cuisine. And a lot of it looked good. I just wasn't quite sure what any of it was. I dug in and asked what each dish was in turn. The Peking fried duck was amazing. There were rice bowls – a local delicacy. There was a beef dish that reminded me of carpaccio until I realised I needed to cook it in hot water. I tried ribs, which were good, frogs' legs – never again – chilli beef, and mountains of dumplings. I deployed my best poker face when I needed to and pressed on. I wanted to show them I was keen to explore their culture. That I wasn't another ignorant Westerner who thought people would understand me if only I shouted louder.

I could see what looked like onion rings in front of me. I knew they weren't onion rings, but that was my only reference point. I told myself it was very probably calamari, which I love, and I picked one up and start eating. My first thought was that it was very chewy. My next was that it tasted earthy. It wasn't good. I was chewing hard, when in the corner of my eye I noticed one of the Chinese girls. She was holding a hand over her mouth and laughing. It was the kind of stifled laughter that kicks in when you know you probably shouldn't be laughing. The girl spotted me looking at her and gasped.

'What?' I said, and for an awkward few seconds there was silence. No one replied. And then, they all started laughing.

'What is this?' I said, in between chewing. 'What the hell is this?'

The taste was more horrible with each straining movement of my jaw. When I thought I could get away with it without choking, I stopped chewing and swallowed. To my relief, the 'calamari' went down without the need for the Heimlich manoeuvre.

The whole table were in stitches now. 'What *is* that?' I said, again. It took me a few more minutes of persuasion but eventually someone passed me their phone. I stared at the screen and a photograph of a cow's anus. Why? Just why? Dwayne's head shot forward, and I realised he was gagging. He was on his feet heading through the door to the balcony outside. He was pale and he was clutching his mouth. Frederik – a Dane who was in town for a job interview with KVK and had had a lot to drink – picked up one of the crispy rings and stuck his tongue through it, making the rest of the table laugh. Dwayne appeared to have been sick in his mouth.

I was in China and I'd eaten asshole, and it tasted like shit. Literally. After dinner, we wandered the streets of downtown Ningbo. Buildings loomed above us, lit in matching, changing colours. Fairy-lit trees lined the streets, and in the distance beyond a fountain, what looked like a giant toy train set rolled past, with people on board.

Later, Ashley phoned. There was a seven-hour time difference. It was one in the morning there and I was in a bar the Danish KVK guys kept coming to because the owners imported their own Heineken. I overheard one of the staff say, 'that's KVK', like they were part of the furniture in there. It's open-fronted and felt like a cross between an American sports bar, a UK coffee shop and a Danish pub. Around us, young people sat at tables, smoking. I was eating a pizza and sipping an ice-cold beer.

'What are you up to?' Ashley said.

'I'm just getting dinner,' I said.

'I thought you got dinner earlier on?' I could hear the confusion in her voice.

'Er … Not really. I'm getting it now,' I said, trying not to think about the food I'd abandoned earlier.

After the press conference, other news agencies tried to get in touch, but we had a schedule to keep. Shanghai was next. My room was on the fifty-second floor of a Marriot hotel that towered over the Yangtze river. I was unused to the scale of it all. Where I come from, landscapes and the sea are imposing, but the buildings sit lightly in their surroundings, as though they have sprouted from the earth and evolved that way. From our rooftop bar, I could see the lights of Shanghai harbour stretch out in front of me, and it was glorious. I had never seen a view like this.

The following day, the rain was torrential. I walked the city streets in jeans and a T-shirt, soaked to the skin. I tried to hail a taxi, and I realised every single driver was blowing me off. I didn't blame them. I was a white guy who probably couldn't speak the language, and I wouldn't be able to describe where I wanted to go. I carried on walking, inhaling and absorbing as much of the city as I could. I knew I'd be back, and I was looking forward to it already.

I'd been away for a few days, so contractually, something needed to have gone wrong. I got a head start on that when I was sitting in the KVK office, catching up with Palle.

'How are the guys doing?' Palle said.

'Fine,' I said. 'They're fantastic.' Craig and Graeme were keeping things going and Jack was out with them, bulking up the numbers to keep everything smooth. Jack helps us out part-time, and I was in the middle of telling Palle that the team are really conscientious, that they care about the equipment, and that nothing at all had gone wrong. I would have had a phone call if it had. I wasn't even worried when I realised there was a missed video call on my phone a few seconds after we had stopped talking and the girls in the office handed us beers. They were always handing us beers. As soon as we stopped talking or stood in one place for too long, one of the girls approached with a bottle. We were walking out of the sliding doors at the front of the office, Palle four steps in front of me, when my phone rang again. It was Craig.

'I'm really sorry. Something's gone wrong, and it wasn't my fault.'

'Wait, wait, wait,' I said. I turned to Palle. 'What was I just saying?'

'You were just telling me how conscientious the guys are,' Palle said to the phone.

I took a long swig of my beer. 'What's happened, Craig?'

'We strapped the crush down to the trailer, but Graeme forgot it was strapped down and he lifted it.'

The KVK crush functions on an elevator design. It lifts the cow off the ground, with hydraulic rams. The rams can shift a two-ton bull. It seems 'Croatian' Graeme raised the crush, forgetting about the straps, and watched – in horror, I'm guessing – as they ripped clean through the horizontal bars on the sides. Craig was visibly gutted, but it wasn't his fault.

Later the same day, I was with Palle, when my phone rang again. Craig's face loomed large on the screen. 'I'm so sorry. I didn't mean to.'

'What happened?'

'I ran over one of the kids' go-karts,' Craig was not having a good day and looked like someone had knocked the wind out of him.

Thankfully the return journey was the planned nineteen hours, with only a two-hour layover in Dubai. When I got home, shattered, I told Craig and Graeme I was glad no one got hurt and that I was grateful they told me what had gone wrong because it saved me the bother of looking. I was just relieved to be back home, in Scotland with Ashley and the kids, enjoying everything that June had to offer.

Chapter Eight:
Nostalgic for Now

My Granny Parker liked to say that the best day of the year was the twelfth of July. According to Granny, statistically speaking, the twelfth was the most likely to bring sunshine. I'm not sure where she got that info, or even if it's true, but as a factoid it's lived in my head, for years. July brings to mind long, sunny days and the seven-week school holidays we had each summer, as kids. It makes me think of working on the farm with my father, golfing with my brother, and trying to play cricket with the whole family.

In my freeze branding days, I carried dry-ice – frozen carbon dioxide – in my van. On the hotter days, I'd treat myself to boxes of Wall's Solero: vanilla ice cream, wrapped in a layer of fruit sorbet. My ices would chill among the minus-seventy-eight-degree blocks, in the back of my Citroën Berlingo van. My brother James – a sculptor – had sold me the Berlingo for £400, with most of his signage still on the side, reading 'James Parker Sculpt'. I still think he got a good deal out of that. The ices were usually soft by the time I got them in there, but they came out solid and took half an hour to thaw enough to be bitten into without a visit to the dentist. For me, there was nothing quite like taking a break from all that effort, sitting on a drystone dyke, with the warm sun on my skin and that sweet, citrus flavour on my tongue.

I read somewhere recently that we should try to be nostalgic for 'now'. I love that idea. It's a powerful sentiment. As I write this, I'm forty-two years old. If I'm lucky, I'll be looking at pictures of 'now' when I'm eighty-two. I'm certain I'll feel that familiar warm glow somewhere in my chest, like the one I get today, when I look at pictures of the kids when they were babies and toddlers, and on their first days at school. I won't be thinking about the day-to-day stresses, the dental appointments or the tax returns. Memory will filter the background noise and the silly arguments of the here and now. All I'll see is how lucky I was, how lucky I am, right now.

You can look at an old picture of yourself; a younger you, sitting in the sun with your brother or sister. It probably wasn't a perfect day. If you live in Scotland, the sun might only have come out five minutes before the picture. Maybe you'd just had an argument with your brother. Maybe you had a filthy migraine. Maybe you realised, even then, how embarrassing that homestead haircut would one day be. But if you're like me, you won't remember any of that. So why wait? Wouldn't we all be better off if we could just tune out some of the pretty distractions and appreciate the bigger picture of the here and now. I realise it's just another way of saying we should all live in the moment, but I love the sentiment so much I think it might become another tattoo.

July is a manic point on the farming calendar. Growing up on a mixed beef and sheep farm meant there was always some way I could help my dad. It was shearing season. I looked forward to that every year. The shearers – a team of athletes, each of whom could strip the fleece from a sheep in under a minute – would arrive on the farm with their trailer, their music and their banter. Living on a farm can mean you don't see many people. The occasional influx of outside help kept things interesting. The shearers are respected for their work ethic. It's generally a young person's job – although I know a couple of old dogs who are still at it. It's hard work; hard on the back and the joints and everything else. It's hot, and you sweat, and the waxy lanolin that sheep produce to protect their wool seeps onto your skin, stinging any cuts it hits.

The shearers wrestle one sheep after another from a race on their trailer, onto a shearing board, and then, through a succession of turns and blows of their electric clippers, like some mad dance, they remove the fleece in one tidy piece. I was in awe of them, with their lean, muscular physiques, their tanned skin and their stories. They wore singlets and shearing trews, so they could move with ease, and moccasins, so they could slip across the wooden boards without the restriction of soles. Best of all they travelled, and not just to other farms. Every year, they'd follow the sun round the world. They'd shear in Australia and New Zealand, then come home for the start of the season here, earning good money in the process. They seemed to me like a motley crew of pirates. What a life. These days I have a bit of that myself. I might not get to Oz, but I get to meet and work with different people, and that only makes the job more interesting.

Shearers are paid per sheep, so speed is a priority and it's a matter of pride. There are competitions all over the world, with marks awarded for speed and quality, and they make it look easy. Once they were done with a sheep, they'd kick the fleece backwards, towards us kids, and we'd scramble to get it. The task of rolling fleeces was ours. We'd fold in the sides, then roll them up like sleeping bags and tuck the ends back into the centre to tie everything into a tight ball. I suppose that was a matter of pride, too. The land we lived on was scattered with whin bushes. Their needle-like projections would catch on sheep as they passed by and stick to the fleece. Inevitably, they'd find their way into our young fingers at shearing time, just to add to that lanolin sting. Here in Southwest Scotland, sheep are bred for meat. Their fleeces are coarse. They're good for hard wearing fabrics, like tweed, but you wouldn't want to wear a jumper made from our wool.

It was a hard day's work, even for the adults, but I loved it, and in some ways, I still miss it. We'd pile those fleeces into huge, grey woven sacks, and when they were full, Mum would sew them shut with the biggest curved needle I've ever seen, then tie on a brown ticket to say where the sack of wool had come from. We'd drag the bags through to a rusty old hay shed at the entrance to the farm, and stack them like a giant running tally of all our efforts. You might think that sounds like a profitable day, but the wool industry has been in decline in Scotland since the fifties, and every year, farmers shear their sheep at a loss. It's only done to keep the sheep clean and comfortable in the height of summer.

July is a fantastic month for trimming hooves. It's a time for farmers to 'reset' their herds; when the moisture levels drop, cows are outside on dry ground and feet can heal. The boys are off school for the summer and that means I can take them to work – just one of the freedoms I love about working for myself. For Keir and Campbell, helping out is a massive learning opportunity. I have no idea what they'll go on to do in life, but everything they learn here is transferrable. It's not about hoof care. It's about building confidence, resilience and common sense. It gives them a work ethic, and the kind of attitude it takes to run a successful business. It shows them that sometimes you have to get your hands dirty.

I took Keir along with me, and Craig, to a farm about twelve miles from Laigh Kirkland. It backs onto Barmeal – the farm I still think of as home. The family have been customers of mine for fourteen years now. They are the quintessential farming family. Everyone mucks in and helps out, and the farm is the centre of family life. Keir helped us get the cows down the race and into the crush. It wasn't easy but it was a good opportunity for him to push his boundaries and build his confidence. Safety comes first. In the past, I've taught Keir to stay away from cows. Now he's a little older, I'm encouraging him to get involved and I'm teaching him how to move them safely. He's not allowed in with them, just yet. He needs to learn just how powerful these animals are, how bad it can be if you're in the wrong place at the wrong time and how to balance that with the need for speed and efficiency in the working day. For now, if he needs a break, he only has to ask.

I'm more of a carrot than a stick man. He's cautious and I'm happy with that. His little brother doesn't have the same level of sense, just yet, but we'll get there.

Growing up on a farm gave me a lot of freedoms kids from towns and cities didn't have. We might not have been able to do whatever we wanted. We couldn't go to a leisure centre or tenpin bowling or to McDonald's whenever we felt like it, but we had the freedom to explore work in a way many adults never do. We had the chance to learn by our mistakes and to understand how to overcome those hiccups and move on from them. It was a slice of grown-up life, and when you're a kid, that's always a big deal. That's something I've always wanted for my own kids.

Keir, Craig and I were getting through the cows, quite the thing, when the farmer's son appeared with a tell-tale shiny package. Plastic boxes topped with aluminium foil. Fresh bacon rolls. They were a bribe from the farmer, in exchange for appearing slightly earlier than normal. It was 8 a.m. but at this time of year the sun was already sitting high in the eastern sky. I watched Keir enjoying his bacon roll. I could see a younger me, keen to earn his place in the crew, which, I'm proud to say, Keir has.

It took me a few wrong turns to find my own place. My brother, James, joined the Royal Air Force in 1999. To my eyes, back then, he had it made. His life had structure and discipline. He travelled the world. He always had cash in his pocket. He looked sharp. To someone like me – restless and drifting, always chasing the next night out – that kind of stability was enticing. I craved the sense of belonging and purpose; the money, and travel wouldn't go amiss either. So, I decided I needed to follow him in. James was horrified, and he didn't sugarcoat it.

'Graeme, you'll hate it,' he said. 'The RAF's nothing but red tape and health-and-safety bollocks.'

If you knew me back then, you'd know the effect that kind of advice had. I was only more determined. I had spent the previous couple of years studying photography, at the George Street School of Art in Dumfries. I still lived there. The nearest armed forces recruitment office was in Carlisle, thirty-five miles away, just south of the border. I made my way to a grey-fronted building in the city centre and spoke to a flight sergeant on the recruitment desk. Everything seemed to happen quickly after that. I was thrown into a battery of assessments. There was a bleep test, an IQ test, a full medical and hand–eye coordination drills. Somehow, I passed everything with high scores.[4]

I had gone in the door intending to join as a photographer after studying Photographic Science at the George Street School of Art and Design in Dumfries, but the man behind the desk was having none of it. 'Why not apply as a Physical Training Instructor?' he said.

---

4. If you read my first book, *Bruised Sole*, you'll have noticed that I managed to forget to mention this entirely! Suffice to say I tried so many different career pathways between school and hoof trimming, that I sometimes miss a piece of the jigsaw puzzle. Plus, it's still something James likes to take the piss out of me for, so naturally I prefer to give him fewer opportunities to do that.

It was a better entry point, so I re-took the tests and scored even higher, but the flight sergeant wasn't done with me yet. He dangled something shinier. Why didn't I try for entry as an Aircrew NCO. I could be a loadmaster – the person who calculates the weight and balance of an aircraft, so the thing flies straight. That was a much higher rate of pay. I felt sure the discipline would straighten me out. It sounded like the life I needed. The only big downside was that the training was at RAF Cranwell, in Lincolnshire, nearly eight hours away by train.

By the time I went to Cranwell for further assessments in 2003, I'd been back and forth in the recruitment process for a good eighteen months. I'd kept myself busy, working in bars and nightclubs in various places, but really, I was just treading water and waiting. James kept muttering that I was wasting my time, but to me, each promotion up the entry ladder felt like a vindication; proof I was doing the right thing. So, I packed my bag and off I went. Cranwell was immense. It's more of a college than a military base, with dormitories, gyms, and lecture halls. I can still picture the brick buildings towering around parade squares.

We were collected from the station in a little white minibus. It must have held around sixteen people. It didn't take me long to realise everyone else looked smarter than me, better dressed, better pressed, more at ease. I was the scruffy outsider. A poor relation. The driver banged on about the brilliant, young, Cristiano Ronaldo, all the way there. He insisted that anyone could be that good if they just put their mind to it. I remember thinking it was one of the daftest things I'd ever heard. Some things just couldn't be faked.

The days that followed were a blur of tests. Teamwork drills. Fitness trials. More tests. Interviews. To be aircrew, you had to be the fittest of the fit. You had to be smart, articulate, able to fit into a team. I did well. That first evening, we hit the on-site bar. Nobody drank heavily. We all knew we had tests to get through. For the first time in years, I felt like I was part of something. Then came the final interview. We sat in a cavernous hall, chairs laid out in neat rows, like exam time in high school.

We had been told to read as much as possible about current affairs and geopolitics. We bounced with nervous energy. We peppered each other with questions. I knew most of the answers. I could feel the team spirit in the room. When my turn came, I perched on a comically small chair in a vast room. Three officers loomed over me, like I was a specimen to be measured and categorised. They threw questions at me. What did I know about the current situation in Sierra Leone? What were my big life experiences? I held steady. I spoke clearly. When I didn't know the answer, I said so. At the end, I stood, shook their hands, and left, feeling strangely proud. For once, I hadn't bottled it.

Hours later, I was summoned to another room. The sign on the door read 'Review Officer'. My stomach dropped. Inside sat a senior officer, with greying hair, his polished uniform bristling with medals. The kind of man so high up no one dares to challenge his waistline. He opened my file and flicked through it. Then he looked at me, considering something.

'I'm so sorry,' the officer said. 'This is ridiculous.'

I froze. What had I done?

The man shook his head. 'The committee says your interview was perfect. But they've failed you on enunciation.'

'My ... pronunciation?' I said.

'Yes. The way you speak. It's not clear enough.'

And that was that. Eighteen months of testing, prodding, having my expectations raised, all ended by my accent. I may not have been as broadly spoken then as Cameraman Graeme is now, but I'd always sounded broader than James, and clearly my time in Dumfries hadn't softened my Shire tones enough for me to pass muster. I suppose it's hard enough to make out what someone is saying on a radio, without adding in my quick, Scottish tongue, but it was a tough pill to swallow. The officer apologised again and gave me a P3 entry, so I could reapply the following year, but inside, I knew I was done. James had been right all along. That night, I drowned my disappointment in Guinness at the on-site bar. We bowled again, on two little alleys, tucked away in the corner, egging each other on. A speedometer flashed up our fastest throws. We laughed and drank and pretended none of it mattered.

The next morning, hungover and hollow, I boarded the train north. I found my way to carriage F, seat nineteen. It was a table seat by the window, at least. The two seats across from me filled quickly. Another man dropped in beside me and nodded off straight away. My head thumped. My stomach churned. Guinness sat inside me like black tar. As the train rattled and swayed, I could feel the pressure building in my gut. I felt myself clenching and prayed it would pass. Once or twice, I let small ones slip. I made a show of glancing at the man sleeping next to me. I wrinkled my nose, in case anyone watching thought the smell was coming from me.

By Newcastle, I was in trouble. The pressure was unbearable. Every muscle in my body felt tight with the effort of holding back. *Just hold it*, I told myself, *just hold it until Carlisle*. And then I had a thought. *Maybe … just one more silent one. One tiny release.* I tilted myself onto one buttock, relaxed my muscles, just enough, and in that moment, you might say the bottom fell out of my world. I sat there, frozen. Panic screamed in my skull. My guts gurgled, and a molten wave surged into my boxers. Then came the smell, so thick you could choke on it. I could see the reaction on the faces around me. My heart raced. My ears burned, but I sat there, rigid, every part of me held in place by fear, except for my head, which – and I'm not proud of this, really, I'm not – tilted in the direction of the sleeping man. I pinched my nose in disgust, but there I stayed, in my own mess for over an hour, terrified to move in case it ran down my legs.

When we reached Carlisle, I bolted. I could feel it clinging, burning, semi-dry, against my thighs. I moved as quickly as I could without dislodging anything. I found a branch of Marks & Spencer, grabbed a cheap pair of black trousers, a pack of underwear and two carrier bags, and scurried into the station toilets. Safe in the cubicle, I stripped and wiped until my skin was raw. I stuffed my ruined clothes into one of the green M&S bags and pulled on my clean boxers. Normality, almost, but I didn't want anyone to smell the bag under the door as I changed into the clean trousers. There was a small horizontal window above the toilet. That would have to do. I opened it as far as it would go, then I stuffed the offending bag through the opening and dropped it into car park. It landed with a thud.

'Ah, you dirty bastard!' a man's voice shouted from below.

Oh well, I thought. At least it's gone.

Then, I looked down and my stomach dropped again. I had thrown out the wrong bloody bag. My brand-new trousers were in the car park. The filthy ones lay in front of me inside the other bag in the cubicle. I checked my watch. My train was due. There was no time to fix it; no time to fetch my new trousers or buy another pair, so I did the only thing I could. I stepped back into my soiled trousers and pulled them up. The filth smeared its way across my thighs again. I left that toilet stinking, ashamed, but oddly comforted by the fact that my boxers were clean.

That wasn't my proudest moment, but it wasn't my worst. Not even close. Sometimes the moments you want to re-member least end up shaping you the most. That day taught me about pride, about shame, and about listening to James once in a while. It taught me something else too. No matter how far you travel, how grand the uniform, or how shiny the dream, you're never more than one bad Guinness away from being humbled.

Chapter Nine:
A Turnip for the Books

In the far-flung reaches of Southwest Scotland, where agriculture is everything, we dance to the tune of the weather and the rhythm of the seasons. The first Wednesday of August is ALWAYS reserved for Wigtown Show. It feels like the high point of the summer sun. It's arguably the highlight of the year for a lot of people. The show is a vibrant gathering of everything and everyone agricultural in The Shire. The rugby ground at Bladnoch fills with local businesses: feed suppliers, equipment manufacturers and tractor dealers, all out in force, talking to customers, old and new, usually with the enticement of booze. Getting drunk is free and easy on show day. There are displays of showjumping, motorbike stunts and monster trucks. There are animals – cattle, sheep, horses, donkeys, goats and poultry, all paraded and judged in the quest for prized rosettes, handed out by celebrities. I remember one particular Scottish footballer who got royally drunk before dishing out those ribbons.

As the days dwindle in the wake of the show, locals are frequently heard to say that the nights are 'fair drawing in', like some Game of Thrones-esque reminder that winter is indeed coming, even if it is a fair way off yet. In the depths of winter, the light is captivating. The sun sits low in the sky, casting shadows and throwing colours that don't appear the rest of the year, but it only hangs around for seven hours and if I didn't stay positive, I might say it's depressing. So, August feels like a last chance to get things done. The days are long, and that prima donna of a yellow ball is still in the sky.

When I was thirteen or fourteen, I got my first real job away from the family farm. I worked on a nearby farm for a local legend. Stuart was a mountain of a man, in a tweed cap. A gruff character who took an aperitif in the afternoons and arranged carnival events for the locals in the summer. When we were younger, he'd stick crash helmets on our heads and drive us up and down the farm road on one of his scramblers. I can still hear the ripping sound from the engine and feel the rush of wind on my face. He'd plant trees on his farm, and when people asked him why, he'd say they were for his descendants. In my experience, that's the kind of long-term thinking the best farmers have. The land is only ever in trust for future generations. Back then, I thought Stuart was seriously well-to-do. He had a huge farmhouse that had been in his family for generations, and he always seemed to be throwing parties. Now I realise he was – and is – just a really decent guy. Despite my age, he treated me like a seasoned farm worker, and that meant a lot.

Braefoot is the place I learned to drive a tractor, properly. Don't get me wrong. I'd had a go at The Knock, but only ever when no one was looking. Here I had no choice but to use them like it was no big deal. It was a very big deal to me. I felt useful, like I was actually helping, and not just a token presence, looking for things to do and getting in the way. I'd say that was my first real taste of adulthood. I had to work hard. I was responsible for something. You might think I was young to be working the land and driving tractors, but here in Galloway, if anything, I felt old. There's no legal age to drive a tractor in a field and it was all I wanted to do.

Stuart had a couple of hundred cattle and a few hundred sheep. From his farm, I could see the Fell of Barhullion, with its crowning cairn. I'd been looking up at that hill for years. We'd climbed it once, as a family, looking for the remains of the fort on the top. It was a moody kind of day, and lightning struck the loch as we walked past it on the way up. Now I was viewing it from a different angle, as a tractorman.

I got into the daily rhythms at Braefoot and explored the unfamiliar land, a place I'd stared at from the school bus or seen in the distance but never really appreciated. Every day was a wee adventure. Our land, with its rich soil and mild climate, is perfect for growing certain crops. Stuart knew what he was doing and took full advantage of what he had. He cultivated acres of turnips to feed his sheep. Harvesting these neeps – as we call them – was my job. I had a Ford tractor that must have dated back to the seventies or eighties. The chipped blue paint, rusting white wheels and smoking black chimney might not have looked much, but to me she was magnificent, a great mechanical beast, and a symbol of my newfound responsibility. I'd watched Dad drive tractors all my life. Now it was my turn. Every day I'd climb aboard, full of purpose, and drive to the steading to pick up a trailer. The trailer was a relic, maybe even older than the Ford, all wood, metal and smooth tires. The years had coated it in soil, leaving it a monochromatic brown, but it was solid, with low sides. Perfect for loading my neeps. I'd hitch it on and bump along one of a network of rutted tracks, my body swaying with the lie of the land, as the seat beneath me creaked and groaned, over the diesel rattle of the tractor's engine. The sun streamed through the dusty windows of the cab, coating me in a warm glow. I don't remember feeling cold or wet on those days. I'm sure it happened at some point, but my memory is of the sunshine and the satisfaction of a hard day's work.

The Ford had a hand throttle – a blue stalk, next to the thin, black steering wheel – and a clutch pedal that didn't seem to be connected to anything. I'd push the throttle away from me, wind up the revs and jam my foot down on the clutch. I'd grind the stick to find the right gear, then lift my foot, re-engage the clutch, and hold on, as the tractor lurched forward, into another day of growing up.

Day after day, I traversed the farthest reaches of the farm. Beyond an old wooden gate lay a vast, rolling hill, the furrows where the turnips grew etched into the dark soil. This was Shug's domain. Shug Calder was the main worker on the farm, a tall, wiry man with a buzz cut and a quiet wisdom. At the time he was probably only about thirty, now I think about it, but I thought he had a few more miles on the clock. Shug's brothers worked on our farm when I was small, so I felt at home with him. We trim at a farm nearby once a fortnight, now. Shug works there and more often than not he'll stick his head over the wall and ask how our day is going. He's a welcome sight; a friendly face when we're up against it. Shug had already been over the field with a set of cutting disks, so now the turnips lay on their sides, sliced from their earthy roots.

My job was to load the trailer with as many neeps as it would take. Nowadays it would be automated, but back then, in the mid-nineties, we had a more inventive way of doing things. I'd line up the Ford at the bottom of the hill, so the two front wheels straddled the furrows. I'd set the hand throttle low and crunch my way into first gear. As the tractor moved off at a crawl, I'd swing the door open and jump from the cab, and – Elon Musk, take note – the furrows would guide the small, ribbed front wheels back up the hill as the back wheels did the work. Health and Safety wouldn't have loved us, but that runaway blue tractor was really only walking.

I had an ancient pitchfork, with two rusty, curved prongs, and I'd use it to fill the trailer as quickly as I could. I'd stack those turnips high, then I'd catch the tractor, climb aboard and drive my cargo to whichever field Stuart had specified. When I got there, I'd pull a stiff lever, activating a hydraulic ram to tip the trailer, and I'd drive along at the same strolling pace, serving up breakfast for the sheep. They'd rush in, crunching through turnips, like they hadn't been fed in days. Then, I'd go back to my hill, moving trailer load after trailer load, for as long as I could.

My forking technique was solid. I'd shove my left prong through one turnip, leaving just enough room for the second, then I'd swing my whole body around and bang the shoulder of the fork on the side of the trailer to break them loose. The turnips sat in uneven rows, and I was able to build a rhythm. Stab one turnip, stab a second, bang the fork on the trailer, turn back around. Stab again.

I could happily do this for hours. Writing it all down, I realise that probably sounds a bit boring, but it wasn't. I was in my own little world, with my own responsibilities, and I loved it. There were only two rules: keep up with the pace of the tractor, following those furrows, and get the job done. Sometimes a front wheel might hit a stray turnip, and the tractor would bounce sideways into the next furrow. That always woke me up a bit, but that's how we rolled back then.

Even in the dry summer months, the ground was muddy, and the claggy soil would attach itself to my boots, so my feet got heavier as the day went on. My hands got layered with a thick, grey crust from all the times my fork banging technique failed me and I had to liberate a turnip by hand. I remember thinking I must be getting fit, that I must surely be growing muscles. I was keen to impress the local girls. I don't think it worked.

Once Stuart was happy that the sheep were full, I'd start delivering my turnips to a shed, back in the steading, piling them as high as I could. My top score was fourteen trailer loads in a day. It's a record I'm still proud of, even if I never told anyone until now. Those days on Braefoot taught me a lot about hard work. They showed me just how far I could push myself, physically, without anyone watching over me or cracking the whip. They taught me that hard work pays off, and that there is a satisfaction in over-delivering on other people's expectations. Stuart taught me the value of treating everyone like an equal.

It was thirsty work, but turnips are sweet. I doubt many people reading this have tasted turnip freshly cut from the ground, but back then I had a good supply. I'd pick one up with a grubby hand and smash it on a rock to expose the bright orange flesh at the centre. It was sweet and succulent, and it quenched my thirst on those long, summer days. I can still taste it now. I've tried to recreate the effect, but when I buy one from a shop today, it just isn't the same. It's probably a few weeks old and it doesn't come with the relief of quenching a well-earned thirst.

August, now, is very different from all those years ago. I'm usually busy on a small tractor mower, keeping our couple of acres of lawn in check. Five-year-old me would be very proud. *Thirty-five*-year-old me would be proud. At times, I didn't think I'd end up here, in a place like Laigh Kirkland. I dreamt of it, but I struggled to believe it was possible for a long time. Hoof trimming changed all that, but during the early days, when Keir was a baby and we lived in a three up, two down housing association place in Kirkcowan, I spent my weekends fixing up my rapidly failing equipment. Everything was already old when I bought it and long days and heavy, stubborn animals took their toll on the crush and the operator.

At this time of year, I can just get on with my job. In other months, things get in the way. I'm battling the elements, the shorter days and those damp feet. In August I have no weather or moisture concerns. The mud dries out and falls off hooves, leaving them cleaner and more visible, allowing more precise cuts with every stroke of my knife.

The eighth of August 1988 was quite a day, even if I knew nothing about it at the time. In fairness, I wasn't quite six then. Eight, eight, eighty-eight – Ashley's birthday. I like to tell anyone who'll listen that someone up there decided Ash might struggle to remember anything more numerically complex. Luckily for me, she knows I'm only joking when I say that. Ashley is a bit of a cleaning enthusiast. I sometimes worry she'll tidy me away if I stand still for long enough. So that I don't smell too much of cow manure on her birthday, Ashley, Keir, Campbell and I pack our cases and head somewhere hot. Usually, that means southern Spain or one of the islands in the Mediterranean. We love spending time together, just the four of us, but sometimes we take family along for the ride. A couple of years ago, we were driving through one of our favourite resorts, Puerto Banus, when I got a phone call.

Puerto Banus is a white-walled tourist-town, nestled on Spain's southern coast. Originally it was just a small harbour. Now there are yachts and a marina in that harbour, supercars on every corner, and exclusive eateries all along the water-front. We usually rent a villa. My ever-busy head won't let me relax in a hotel environment. We have young children. They make a lot of noise. I don't want to worry that we're disturbing people, but it is the kids' holiday too. They should be able to go a little crazy. In a villa, away from people, they – okay, *we* – can go nuts, without the worry of ruining anyone else's holiday. Thinking about it, we probably wouldn't disturb other people, but reality gets distorted when my anxiety is in full flow, and the villa just means I can relax and enjoy my holiday.

Anyway, as usual, I've wandered off on a tangent. We were in Puerto Banus, driving down the central avenue, in our rented SUV, sun shining, Ash sitting beside me with her feet on the dashboard, when my phone lit up with Craig's face. That would be a shock for anyone, but this was 2021 and it was rare for me to be away with hoof trimming ongoing. Going away and letting Craig get on with it was a giant leap at that stage. Craig's an excellent hoof trimmer, but I was away, so his face on my phone cranked my pulse up a few beats. He wasn't completely alone then. He had Kevin with him. Kevin is Ashley's brother. He didn't work for us at the time, but he'd generously offered to help Craig for the week, getting the cows into the crush and updating the records. Kevin's a highly experienced dairyman, so between them they had the cows covered. What could really go wrong?

I decided to ignore the call. I'd phone Frodo Craggins back later. Work could wait but this moment was fleeting. I wanted to enjoy it. But when I silenced my phone, it lit right up again. I declined the call again. It lit up again. *Oh shit.*

'Hi Craig. Are you all right?'

'No … No, I'm really, really not okay.' I could hear the gravel in his voice. He was talking through tears.

Back home, it was a Thursday morning. Craig and Kevin

had just wrapped up some filming. I'd asked Craig to record the action while I was away, and the pair of them sat on a big straw bale, outside the vast shed they'd been working in all morning, tea in hand. Craig held the camera at arm's length, shoehorning them both into shot. The sky was a pale blue. The birds were singing. It was a perfect August day. Craig had sat Kevin down for a review of their week. Kevin's not a big talker so that didn't take long. 'So far so good,' was his reasonable conclusion. Fate has a cruel sense of humour.

Filming done, Kevin drove the pick-up and our Appleton Steel crush to the edge of the shed. Craig washed the crush down with a power hose. That's standard practice when we leave a farm. I lived through the foot and mouth disease epidemic of 2001. I saw farms in lockdown and fields blazing with cattle pyres. That image, and the smell, have never left me. Hygiene is how we stop it happening again. Once he had the crush all shiny and clean, Craig took the wheel. He drove through the doors at the end of the shed, out onto a ramp and into a viral fail.

He should have taken a right turn down a gentle slope, through the farmyard and out to the open road. He didn't know about the gap on his left-hand side, or the CCTV covering the whole scene. He zigged when he should have zagged and turned slightly left. In the resulting grainy footage, you can see the chunky front left wheel of the pick-up bounce off the edge of the ramp, followed by the back left tyre. Craig would later say the whole thing happened in slow motion, but on camera, it's pretty quick.

The front wheel went down the ramp and over the shallow edge at the bottom. No big deal. Neither Craig nor Kevin felt much. But they were on a ramp and the back wheel found the edge further up. The abrupt loss of height told Craig something was wrong. He applied the brakes, just as the wheel of the crush left the ramp, over a two-foot drop. The extra height, the spring in the off-road tyres, the braking force and the momentum of the three-ton crush, combined to send the whole thing into a roll, and with a grinding crash, the mighty Appleton Steel hit the concrete side on.

In the black and white video, captured by the farm's CCTV system, Craig can be seen running back from the pick-up, trying to figure out what had just happened. When he gets to the crush, he stops and throws his arms up round his head. It's like a comic skit, but Craig wasn't laughing. He'd been left in charge for a week, and everything had been going brilliantly. He'd just said as much in his video. Now, he felt like he'd failed. I can only imagine how hard picking up that phone must have been.

'Is Kevin safe?' I said.

Craig's voice was shaky. 'Yes.'

'Are you safe?'

'Yes.'

'Don't worry then,' I said. 'Nothing else matters, whatever it is can be fixed.'

'I've rolled the crush, and it's fucked.'

'Ah shit!'

To anyone overhearing our conversation, the way we talked and the language we used might seem a bit coarse. I mean, it *is* a bit coarse; certainly not what one would expect in polite company, but where we live, it's standard diction. It's an expression of magnitude, and not always in a negative way. In fact, some of the absolute worst words can be used as terms of endearment in The Shire.

A quarter of an hour before Craig phoned, I'd received a text message from a customer back home. He wanted – for reasons which then became clear – to know if I was buying a crush while I was away. 'No, why?' I'd typed back. I knew it was an attempt at humour, but I didn't get the joke; not until Craig phoned and joined the dots for me. Live by the sword, die by the sword, or so they say. And social media was taking a swipe right back at me. Before anyone could say 'viral hit', the footage of the pick-up rolling down the ramp, the sudden drop and the emergency stop, the crush lurching into a cartwheel and, Craig crumpling in defeat – all of it was on every platform you can think of.

In situations like that – rare as they are – I don't get angry. Accidents happen. *Things* can be mended. People, not so much. Craig was upset, but it could have happened to anyone. What he didn't know was that I'd nearly done the same thing a couple of times. From Craig's vantage-point, the gap was hidden. He didn't know what was happening until he saw the crush tipping in his mirror. From a safety point of view, that edge should never have been there.

When I returned from Spain the Appleton Steel needed some love. Hydraulic hoses hung like burst veins, leaking red fluid. The roof looked like a beer can smashed on a drunken forehead. In the end, fixing it took a days' work, but it probably cost Craig more than that in lost sleep. We cut the roof off, reconnected everything and had ourselves a custom, open top version. Everything looked good, which just shows how well built the thing was in the first place.

Unfortunately for Craig, that CCTV video exploded, but the last thing I said to him, before hanging up the phone in Spain, was 'make sure you film everything'.

We made a follow up video of our own, because we wanted to show what really happened, show Craig's side of the story, and just how easy a mistake it was to make. That, and once Craig had calmed down and we knew he was okay, it was too good not to share. With the crush back in shape and the Parkers back from the sun, the weather was getting colder. The kids would soon head back to school and farmers would bring their herds inside. The next couple of months were going to be busy.

Chapter Ten:
The Wild West (Coast)

September's a funny time of year in our part of the world. We get dismal, grey skies, windy days and, just occasionally – when the fiery ball deigns to hang around a bit longer – an Indian summer. We don't typically get much in the way of *actual* summer in Scotland, so we take what we can get and make of the most of it. The days are shrinking, and the kids are back at school. Normal routine takes hold, and they're eyeing up the end of the year and a slow coast through pumpkins, fireworks and Christmas.

Back in the freeze-branding days, this time of year was hectic. In the warmer months, yearling calves are put to grass for the first time. They graze in green fields and bask in the southern Scottish sun when they can. But when autumn arrives, the drop in temperature drives these heifers back inside. Dairy heifers haven't evolved to cope with the harsh realities of our colder climate; they have slight frames, low-fat reserves and short hair. They should be heavily in calf by this point, and farmers turn their minds to herd management. They want their cows to 'calve down' – as we call it – with numbers on their bums. That way, they can identify each one at a glance and follow their progress.

Back in the branding days, I'd go to Campbeltown, on the Kintyre peninsula, every year. I'd drive a hundred and twenty-four miles to the shore of Loch Lomond, on the edge of the Highland boundary fault line, where the Highlands meet the Lowlands, and the mountains rose up in front of me. Then I'd make a giant U-turn, following the road west, between Tarbet and Inverary, then south for seventy-five miles. Yeah. I know. The Scottish concept of distance is a bit strange. In comparison with the US or Canada or Australia, everything is in miniature, but the roads twist and dance in front of you, and it takes time. You see some things, like the Rest and Be Thankful, the highest point on an old military road that runs uphill forever. It's called the Rest and Be Thankful because that's what people did when they got to the top, safely, before the dawn of the car. The road is often blocked by snow, landslides or repair work. When that happens, you have an eighty-mile detour to look forward to. When you do make it through, there's a carpark at the top, with a van that serves the best bacon sandwiches, and no doubt does a roaring trade, with all those road workers.

The drive takes a good five or six hours. That might be a pain, if it wasn't so damn beautiful. The road skirts the shores of sea lochs: Loch Long, Loch Fyne, Campbeltown Loch. At points, you can see your destination, straight across the water, but you have to go the long way round. You have no choice but to give in to it, to enjoy the ride and the view. You wind your way south, flanked on one side by the tepid water of the gulf stream, and on the other by scotch firs. You weave your way over bridges that are hundreds of years old, past castles, that seem to have been there for ever. As epic as the drive is, for me, getting to Campbeltown was always a relief.

The northernmost lands of the peninsula are rough and hilly, but the land at its southerly tip is lush and green, with rolling pastures. It's prime cattle country, a lot like home. And just like home, it has an island feel, like it's cut off from the rest of the world. People here know each other, probably better than they should. The place must have made an impression on me over the years. It's the reason we named our youngest son, Campbell.

I'd rock up in Kintyre with a long list of farmers to see, knowing that when I was done, I'd need to board a ferry and sail the two hours to Islay. In the early days, when I first visited Kintyre, I'd stay with one of the farmers I branded for. It was a place Davie, my stepfather, had stayed many times. They were an incredibly welcoming family, and their farmhouse was beautiful. I just always felt like I was in the way. That's more of a reflection on me than on them. Even today, no matter where I find myself, I never really feel like I fit in. My anxiety eventually led me to stay in a small hotel in the town itself.

Campbeltown has some interesting locals. I'm not a massive drinker when I'm at home and working, but when I'm away on my own – which is rare these days – I like to pass my free time in local bars. One evening I was in a pub in Campbeltown and in dire need of a haircut. This must have been before I met Ashley, who, because she used to be a hairdresser and sorts my 'do' out whenever I need it these days, although some would say not often enough, in the case of topknot-gate. The crew round the bar looked like they'd been cemented in place at the feet. I asked them if they knew where I could get a good clipping.

The old man on the end turned round. His eyes were red, his face leathered by the sea. Even his pint looked like it had seen better days. For a moment, I wondered if I should have said anything at all.

'Aye boy,' the man said. 'There's a geed place oo'r there called Number o' Dicks.'

I thought he must've misheard me. Then I wondered what the hell he thought I'd asked. I decided not to pursue it and went for a wander. Eventually, I found a barbershop. As I walked back down the high street to my hotel, with my fresh cut, something caught my eye, and I started to laugh. Laughing, alone, in rural Scotland, can earn you some quizzical glances, but I couldn't hold it in. Across the street, there was a barber shop. When they'd named the place, the owners had gone with the French alternative to the street number, ten. The sign said 'Numero Dix'. The old man's pronunciation *might* have been wee bit off.

Work in Campbeltown didn't always go to plan, like when I arrived on the farm belonging to Morris Glen, one morning. Morris was a bit of a contradiction. He had the corned-beef complexion of a man who spent a lot of time outdoors but spoke with the kind of upper middle-class Scottish accent that suggested he'd attended Hogwarts. He wore the kind of finely checked shirt that posh farmers must order through a specialist catalogue, and peppered his speech with gruff colloquialisms and swear words, like he was trying to fit in.

The sky above me was a clear blue. I had a view of a calm sea, and all seemed right with the world. As well as being a dairy farmer, Mr Glen was also a vet. It was the first time I'd met him, so I was keen to impress. These days, when I trim hooves, I bring my own equipment, including my own crush. Back when I was in the business of stamping numbers on bums, I relied on the farmers' facilities, and some had better than others. In a way, I enjoyed that. I relish the thought of using equipment that has seen generations of hands. Morris had byres, built in the local whinstone, probably a hundred and fifty years ago. Generally, the farms up there are smaller, and the equipment can be a bit less advanced than what we see in The Shire. There are exceptions, but Morris's place wasn't one of them.

I turned up promptly, introduced myself and asked where I should go to get set up. Then, I fired up my little silver Citroën Berlingo, and I pulled my trailer – a rough, rusty custom job that Davie had built – round the corner. There was no fancy crush and no gated race for the cows to move along, just an aging, flimsy contraption, with a headgate. The kind of thing a stiff wind might lift into the Irish Sea. I'm not an equipment snob. I promise. I've made my own crushes, and other people's when I've had to. I'll make do with whatever's going because I want to keep the customers happy. I worked in customer service for so many years I can't shake the need to please.

Looking at Morris's crush, I was torn. I know how power-ful a dairy heifer is. Most of us would run in the opposite direction, if someone placed a cold iron on our backsides, and a nine-hundred-pound running cow gathers some mo-mentum. You need something to hold her back. Something strong. This crush was not strong, but I knew Morris was a vet and I didn't want to question his authority.

'There's no fucking way that crush will handle those heifers,' I said, regardless. Subtlety has never been my strong suit.

'Of course it will. We use this crush all the time without a bother!' Morris said, with the dismissive tone of a man who isn't used to being challenged.

He was the farmer, and they were his heifers. He knew them and cared for them, and I knew I'd be happy to be proven wrong. It was his call. Usually, I'd be first to get stuck in, guiding cows down the race and into the crush, keen to get on with the job. This time, I decided to hang back a little, just to see how Mr Glen got on. The first heifer wasn't going easily. I remember her to this day; white, thick-set, strong. She circled the pen, all toned muscle and reflex. She dived into one corner, then another, as Morris staggered after her. After a bit of time and some particularly ripe language, he managed to push her along the race he'd cobbled together from rusting gates, and on down to his crush. Maybe, I told myself, I was worrying about nothing.

The heifer accelerated. She'd just reached a gallop when she hit the crush. She smashed into the headgate. Her head found the yoke and the gate closed around her shoulders. Crushes are designed to absorb the force these animals generate. The weak points are usually the bolts holding the crush in place, or the concrete, holding the bolts in place. In this case, neither were a problem. There were no bolts, and so, there was nothing to hold the crush. The cow hit the front gate. The back of the crush shot up in the air, and heifer and crush tipped forward. I stood there, mouth agape, as both somersaulted in slow motion, arse over tit, along the concrete in front of me.

Morris scurried forward, trying to find the lever to release the heifer's head from the yoke, while his farmhand made a grab for all that moving metal and tried to hold the crush in place. Luckily, the cow was fine, and she stood up, wearing what was left of the crush like a piece of industrial jewellery. It took everything I had to stop myself shouting, 'I told you!'

The heifer avoided the number on her bum that day, and so did the rest of Mr Glen's cows.

Back home, in Galloway, September is often the last chance farmers have to get their beef cattle's hooves trimmed. One farmer I've been going to for six years wanted me to trim his bulls' feet. This farmer has around twenty huge Charolais bulls and his cattle are allowed to roam freely, ranch style, across a thousand or so acres of rugged land, just across the water from my house, in the eastern part of Galloway known as The Stewartry. The ground is incredibly soft this time of year and we didn't have the option of a nice, concreted shed floor, so we used what he had: wooden pens on the edge of a grass field. Michael, the Polish stockman, has been there for as long as I've been going. Luckily his English is a lot better than my Polish. His sense of humour is matched only by his work ethic, and we get on well. He's a lot younger than I am but is losing his hair rapidly. He always seems to be wearing the same pullover, one that looks like it has been handed down the generations. I think there's more hole than fabric to it now.

In our part of the world, we call wooden pens 'stocks'. I'm not sure why. It conjures up an image of someone – probably me – being pelted with rotting vegetables as punishment for some misdemeanour. Usually, the stocks are built onto the hardest part of the ground in the field. Here, the ground was wet and the mud between the bulls' feet was soft.

We worked together to separate the bulls from the calves and cows. Strictly speaking, it's not what I'm paid to do, but deep down I love moving cattle around. It takes me back to childhood and working with my dad. We'd syphoned off nineteen of the big, white, fluffy bulls when Michael spotted a lame cow. These cows are pretty much left to their own devices, and most of the time that means happy grazing. This kind of farming is as close to nature as it gets but lameness happens in the wild too, and this level of remoteness can mean problems go unnoticed and stack up.

It was a grey, muggy sort of day, and the bulls were behaving well. I worked in circles, round my crush, skating my way through each trim, moving in the direction of the cow in question. I remember feeling proud of the marks I left behind me on the grass. It had been green when we started, but a few bulls on, things were getting messy. When I was a kid, on Barmeal Farm, I'd play in our front garden with my own toy farm. I'd try to rut up the grass, leaving tread marks with my tiny tractor tires, just like the real ones my dad left behind – evidence that I'd been working hard, feeding cattle. I suppose I still view the tracks I leave behind in the same way, as proof of my own labours.

When I looked down the race, I could see the cow more clearly: an ageing, black, Aberdeen Angus. She hung back. She'd spent most of her life in the open and wasn't keen to get up close with any human if she could help it. I climbed the slippery, green spars of the wooden race and shuffled my way towards her. As I closed in on her, I could hear a buzzing sound. Flies. Lots of them. But I couldn't see where they were gathering.

Beef cows usually live for nine or ten years, but this girl was older. I'd guess she was about sixteen or seventeen. Her coat was patchy. Her hair was thin, and her body rumpled, but she was calm. I crouched down behind her and eased my shoulder into her hindquarters to encourage her down the race. Her front feet dug into the grey mud. I pushed so hard, it felt like I was lifting her back end clear of the ground, then the shift in her weight overcame her and she stepped forward. I repeated the process, digging my feet deep into the mud, driving her, inch by inch, towards my crush. When she was there, I locked her head in place to keep her safe, then lifted one of her back feet and reached down inside the crush, so I could secure her front right foot. She was still calm, but now I could tell where the buzzing was coming from.

I looked down at the offending foot. It was covered in blackflies. It was hard to see. Everything was coated in mud, but her hoof seemed to be open. I pushed a rusty lever and the hydraulic motor on the crush whirred into life, winding a rope round a drum and lifting her foot onto a wooden block where it could rest while I worked. What I saw shocked me. I've seen bad hooves in the course of my work, but this was horrific. Almost half her outer claw was missing.

I'm often asked about water, in the comments on my videos. People want to know why I don't hose down feet before a trim. We normally avoid it because we're working with grinders and a crush, and both are electric. Spraying water around isn't safe for me or the cow I'm working with. Running water can be a bit of a luxury, especially in the middle of a field, in places like this. But this cow's foot was a mess; too much of a mess for me to touch up with a grinder or a knife, and too dirty to see what was really going on. I walked back to my pick-up, found a shiny, new, yellow bucket, and filled it in a nearby water trough. As the cold water hit what was left of the hoof, the animal attached to it jumped in shock. Flies scattered. The putrid stench of necrosis hit my nostrils, and I wondered – not for the first time – what it must be like to have part of yourself rotting.

After the clean-up, what was left was angry and raw. The smell was so strong I could taste it, like metal in the mouth following a visit to the dentist. The wound teemed with maggots, maybe a hundred of them, and in the middle, what looked like a grey-brown piece of wood stood proud. You don't get that on a hoof-trimming course. I have a strong stomach, but this was pushing my boundaries.

I took a cloth and, as gently as I could, I started to wipe away the dirt and debris from the claw. The maggots squirmed and wriggled next to my hand. As I moved in closer, I could see the 'wood' inside the hoof capsule was in fact the cow's pedal bone. The smell was worse, now. I was scared to inhale in case it reached the back of my throat and I threw up. I touched the bone. The cow let out a bellow that resonated through my body. I could feel myself tearing up. Just the lightest touch was excruciating for her, but I needed to ease her pain and there was only one way to do that. I took hold of the end of the bone, applied as much pressure as I dared, and it slipped free.

I sprayed the foot with iodine, and from inside the crevice, the maggots emerged. I lifted my curved trimming knife and carried on working, tensing with each touch of this poor cow's foot. I needed to pare away every last piece of decaying, detached hoof horn. It didn't have to be a perfect job, but I had to lose anything that would attract mud, rocks or maggots. When I'd cleared what I could, I picked up my grinder and buffed off the inner claw, as quickly as possible, avoiding any sudden movements. I wanted this cow out of the crush, but I needed to glue a block to the inner claw. I had to keep that infected digit – what was left of it – clear of the ground. In the end, it took a second block on top of the first; anything to keep her from stepping on that claw and feeling the inevitable pain.

I looked at my work. The claw dripped with blood and iodine. My stomach churned. The infected bone was gone. There would be no more flies or larvae. There was no loose horn to catch muck and stones. I soaked the claw in iodine again, to dehydrate any exposed flesh, then I wrapped it all up in a white, biodegradable bandage, to give the horn the best possible chance to regrow.

A cow like that can never be fully healed. She was missing a bone and half a claw, but I hoped I'd made her more comfortable. I've had a few serious injuries in the past. I know the throbbing sensation she must have had in her foot. Touching the bone would have been painful but removing it must have been a huge relief.

I placed her foot back on the ground and watched her leave the crush. I was done for the day and couldn't help but think how lucky I was, getting to drive home under the soothing September sun, pain-free. I would soon be sitting on the couch with a cup of tea and a biscuit, but this cow would be on her feet and feeling every step. It would hurt for a while, but hopefully, in time, that would go.

September can be a slog. At times, you just need to get your head down and get on with it, power through the lull in the year. After the drama with Morris's crush in Campbeltown, I spent a week branding in the area, then made my way thirty or so miles up the coast to Kennacraig and the ferry to Islay. It would be my first time visiting Scotland's fourth largest island, and I was surprised at just how far it is from the mainland. The ferry terminal, a seventies, utilitarian shack of a building with cheap windows and tired weatherboard, gave no clue of what was to come.

The conical Paps of Jura loom large as you sail west. 'Pap' is an old Norse word that literally means 'breast', and Jura has an impressive set – all three of them. But the scenery was lost on me that first trip. I crashed out on a hard, red faux-leather couch on the ferry, getting a rest while I could. Davie knew the island and its people well. He'd called ahead to arrange my week's work and booked me a room at the Lochindaal Hotel, a small, family-run place in Port Charlotte.

My first farm sat high up on a rugged hillside and gave me a better idea of my new surroundings. I had eighty or so young heifers to brand for a friendly family. I don't remember much about the work, but I remember arriving and feeling welcome, and I remember the characters. There were two strapping, twenty-something sons, their father, and an elderly guy called Bob. I don't think he was related to them, but he worked alongside the family. Bob was what you might call a war hero. He was the type of person who knew everyone on the island – he knew their sons and had known their fathers. In his collared shirt, he looked like he'd been seventy-five since he turned twelve. Bob had never served in the military, as far as I know, but he'd been in the wars, and the brothers loved to tell the stories.

They told me about a time he'd been scraping slurry – cow shit – in a farmyard, with an old tractor. Tractors have grown over the years, and farmers often keep an old one around; partly for the nostalgia factor, and partly because the size means they can fit round corners and down passageways modern vehicles just can't. A lot of farmers show their old tractors at events. The Massey Ferguson 135 is a cult classic; a small, red machine, like something a child would draw; no cab or luxuries, or defence against carbon monoxide poisoning, now I think about it, and only a scraper attached. The scraper is just a rubber-trimmed blade for pushing slurry into a pit where the farmer can store it before using it as fertiliser. Nothing is wasted on a farm.

On an otherwise normal day, Bob had been asked to clear the edge of the slurry pit. This involved pushing all the muck over a steep drop and into the pit – or the midden, as we call it in Scotland. That should have taken him about twenty minutes. Everyone else went off to get started on their own list of jobs, but after about an hour, someone noticed Bob was missing. When they went back to the yard, there was no sign of him or the tractor, but there was a muffled noise coming from the midden. When they peered out over the concrete edge, there he was, sitting on top of a lone, spinning tractor wheel, covered in shit. He'd overestimated the distance, overshot the edge of the pit and cartwheeled the vintage tractor down into the mire. Now it was almost fully submerged in a manure soup, and Bob was on the world's worst merry-go-round. He was, remarkably, unhurt.

The brothers told me other stories, always with the man himself in earshot. Bob had been one of the milk-tanker drivers on the island. Each day, he drove around local dairy farms and picked up fresh milk. His career came to an abrupt halt when he lost concentration on a particularly 'technical' stretch of road one day, and the tanker ended up on its side, in a field. I looked across to Bob for confirmation.

'And you know what?' Bob said. 'I never spilled a drop.'

I'll never forget the look of pride on his face, or the speed of the response. That's what I love about the west coast of Scotland. The humour is harsh, the wit is dry, and the stories are a form of currency, paid for in years of toil on the land, in all weathers. In my experience, there's no situation that can't be made better with a laugh. If you watch my videos and read the comments, you'll see that our own jokes can sometimes overwhelm people who aren't used to them. Editing plays a big part in toning it down for a global audience. From the outside, the sense of humour on Islay, and in The Shire, may seem harsh, but it's a product of the environment. As Billy Connolly once said, 'where the work is hard, people are funny'. The elements and the exertion give us an extra layer of calloused skin, and nothing much is off-limits. When you're up against it, and starting to lose hope, someone making a bad joke at your expense can make all the difference. Everyone has different boundaries, but I know a bit of gentle ribbing is what I need when life throws struggles my way, and when those struggles are physical, tangible – sometimes life and death, where cows are involved – words are just words, but they can change a perspective, lift a mood and make everything feel just that bit more okay.

When I had freeze branded those eighty-odd heifers, we swapped the cold of the farmyard for a cosy farmhouse with a stunning view of the island and the Atlantic beyond. Davie had given me strict instructions. 'Make sure you hand out the invoices and try to get a cheque before you leave the farm.' I've never liked asking for money, so when the farmer welcomed me into his home and said 'go and get the cheque book' to one of his boys, I was relieved. And when he told the other one to 'go and get the beers', my suspicions about the islanders were confirmed. Everything I had been told was true. When in Rome … That autumnal afternoon I sank six bottles of ice-cold beer before I remembered I needed to drive to my bed for the night.

'Ah don't worry,' the farmer said. 'There are no police on the island until Wednesday.'

I was young and stupid, and luckily, there was no one around. Not one of my proudest moments, but definitely not something I would do now.

The Lochindaal Hotel sits on the harbour in Port Charlotte, overlooking the sea loch that cuts into the heart of Islay. Like every other building in the village, the hotel is painted white, but its black window-surrounds and the matching signage mark it out. It's been in the same family for more than a hundred years. Islay is famous for whisky. There are ten distilleries, all with a reputation for their peat-smoked flavour.

It was almost eight when I drove past the now inactive Port Charlotte Distillery, and parked my tiny, French van in the village. It was already dark when I reached the hotel. There was a warm, yellow glow round the door, as I grabbed the brass handle and swung it open.

'Where the fuck have you been?' The voice was screeching, Glaswegian, irritated. I threw a glance over my right shoulder, wondering who had followed me in. Not a soul in sight.

'Aye. You. Where the fuck have you been? I was expecting you hours ago.'

I looked at the barman. He looked back. He cracked a smile. 'Come in. Come in. You must be Graeme. How are you getting on today? You'll be hungry.' He passed me a menu with an out-sized paw.

Two or three locals sat on my side of a dark wooden bar. They didn't seem to notice the exchange. The familiar smell of stale beer and brass polish filled my nostrils. This was not the type of polished setup you'd find in big cities. This was a local boozer with a homely edge; one that hadn't been designed in. The bar was small, with maybe five stools resting along its edge, and one of the best whisky collections I've ever seen, with rare malts nestled among more common names. I say more common names, but what we consider run of the mill in Scotland is considerably rarer elsewhere. When my Grandpa Parker was alive, and living in Yorkshire, any trip south to see him meant taking a bottle of Balvenie DoubleWood, like a tribute. He couldn't get it down there, and he would happily open his present at 10 a.m. to check it had 'travelled well'.

As you can imagine, the island is frequented by whisky lovers – 'smokeheads', they call them, on account of their love for those peaty flavours – but that's in the tourist season. At the tail-end of September, out-of-towners like me are a rarer breed. I stood out like a nun on Tinder. Johnnie, the barman, pointed me to a table at the back of the room. He told me to take a seat and that he'd bring me a drink. I wasn't about to argue with him. I looked over the menu, trying to work out which of the home-cooked meals might be best, and trying not to drool.

A bottle of Miller landed on the table, with a thump and a splash. Johnnie towered over me. 'I don't know what you're looking at that for. It's chicken or nothing.'

I laughed and thanked him for the beer. As I took a cold sip from the sweaty bottle, I wondered if his style of customer service was ever lost on the tourists. Johnnie banged another beer down in front of me.

'What's that for?' I said.

'It's an apology.'

'What for?'

'Your breakfast. If you think I'm getting up at seven, you've got another thing coming. The kitchen's open and there are cornflakes in the cupboard. Help yourself.'

The Lochindaal is a very special place. Maybe the friendliest place I've ever stayed. The conversation I've just described might sound a bit unorthodox, but – trust me on this – if you ever go to Islay, take a detour to Port Charlotte, and go to the Lochindaal. It's quite an experience.

The rest of that week blazed past in a blur of hard work, scenery and stories. As did every other visit in those busy days. Islay is a wild and beautiful place; an isle you drive across on tiny roads that cling to the sides of hills, as you try to avoid the red deer and drink it all in. It's a place of magic, mystery and folklore. Unfortunately, it's a place I don't get to visit anymore. Work just doesn't take me there. But I hope one day to return, and now I'm 'learning' more about whisky, I think it could be educational.

Chapter Eleven:
A Lego Adventure

October in Southwest Scotland is bleak. It's not particularly cold – the average temperature undulates between six and twelve degrees, but we get rained on seventeen out of thirty-one days and we are increasingly experiencing the effects of storms coming across the Atlantic. It feels harsh, wet and windy. Mornings are dank and everything's moist to the touch. The days are shrinking, and the pastures are soddened with mud, so the cows need to be inside. Waterlogged feet are a bad thing for anyone, even hooved animals like cows.

As hoof trimmers, we're busy year-round, but in October we see way more lame cows. All that moisture leads to sand cracks. That probably sounds counterintuitive. You think of sand and automatically picture a desert, but the name's misleading. Sand cracks are caused by mud. Mud saturates cows' feet, then the colder air dries them out, and that leads to a parched river-bed effect. Vertical cracks form in the hoof wall. They're incredibly painful, and their appearance only compounds the feeling that winter is on its way.

In October 2023, KVK Hydra Klov were holding their biennial International Hoof Trimming Conference. KVK is a Danish company, and the conference was in Legoland. Yes.

That Legoland, in Denmark. I've spent a lot of time with Palle and now consider him a friend, so when he asked if I would consider giving an hour-long talk to a few hundred hoof trimmers, I wasn't too sad to escape the grind. As you know by now, I'm not a natural public speaker. I don't live for the rush of a live audience. I'm a bit shy. Okay, I'm a bit scared. Every time I have to do a presentation, my anxiety kicks in again.

This trip wasn't entirely about escaping the Scottish weather. There were possibilities for networking, forging new friendships and rekindling old ones, learning from my peers, and the mentors I've had over the years. There was also alcohol; a lot of it, and that helps. The Hoof GP team are a massive part of the business. I like them to know that, and I like them to enjoy some of the extracurricular opportunities that the YouTube channel brings to our business, so I invited Craig and Kevin along for the ride. Kevin had been working with us as a hoof trimmer for a couple of years and was an integral part of the team. But the boys were playing second and third fiddle as travel companions. This being Legoland, I had to take Mrs HGP.

Scotland only has four main airports carrying international flights: Aberdeen in the north, Prestwick and Glasgow to the west, and Edinburgh in the east. Flights are limited, so the airlines dictate where you fly from. The four of us packed our bags and drove up to the capital. On a cool, crisp morning, we took the short flight from Edinburgh to Billund.

Palle was our ride to the hotel, and as he drove us along a remarkably straight road, I realised something was missing. There was nothing on the horizon.

'How do you know where you are?' I said.

Palle pointed out a small hill, nearby. 'Sky Mountain, the locals call it.'

Møllehøj, it turns out, is Denmark's highest point. It stands at a mere one hundred and seventy-one metres. Mind blown. We arrived in Legoland, and I got over the initial disappointment of discovering everything wasn't made of coloured plastic bricks when I saw the state-of-the-art conference centre. We were shown to princess- and pirate-themed rooms. Kevin was pretty thrown by a giant Lego sword and shield and the blocks on his bedroom walls. Craig was in his happy place, in full-on regression.

I've been to a few hoof-trimming conferences, in different parts of the world. The revolving cast of familiar faces and new people is easily the best part. It's a chance to catch up with friends and with the latest developments in the science and technology of what we all do. The difference now, because of YouTube, is that I'm well-known, and that only adds to the nerves. 'Who does he think he is?' I imagined people saying, as I wandered the hotel. Craig had no such worries. He was determined to be recognised. He'd never been to a conference. This was all brand new for him and I was almost jealous.

We checked in and washed up at the bar. We met people I had worked with, people I'd spoken to online, people I had learned from and still do. We were in good company,

I counted conference goers from Scotland, England, Ireland, America, Canada, Italy, Spain, Denmark, the Netherlands and Germany. Hoof trimmers are a gnarly bunch. We're all different ages, shapes and sizes but we all do a tough job, and we love swapping war stories. Laughter punctuated the background din. If the bar staff had been hoping for a quiet night, they were in for a shock.

Midway through the evening, I went to the bar to top up the drinks. When I got back, Craig bounced up to me, a grin plastered across his face. 'Come and meet my number one fan!'

Craig's not a quiet drinker. He told me he had met a Slovakian hoof trimmer who claimed to be his biggest fan.

We stepped through a door, into an icy Danish night. I couldn't help myself. 'Sorry, did you say one and *only* fan?' I shouted at a slim, blond man.

The man let out a deep, bellow of a laugh. Craig asked him to repeat his original statement. From a distance, he looked like a footballer pleading with a referee.

'I said your one and only fan,' the man agreed. He let out another laugh.

We were all on fine form. Craig was happy, a bouncing ball of sinewy energy. Kevin and I did the rounds, talking to as many people as we could. The drinks flowed, and the anxiety receded.

Next morning, I opened my eyes with a sense of dread. I moved my head a bit, then checked my body to see where it ached. When I was younger, I could party all night. I'd wake up as normal, ready to go again. Now, if I'm lucky, a hangover might last for two days. I didn't have two days. I had a presentation to get through. I pulled back the covers and sat up, waiting for the inevitable grinding behind my eyes. I was shattered, but I seemed to have gotten away with it.

Ashley and I got up, showered, and pieced ourselves together, then took the lift to the breakfast room. I think breakfast might be my favourite part of the whole conference experience. The laddish antics of the hotel bar aren't for everyone. I guessed only a quarter of the attendees had been there last night, but as we strolled through the restaurant, it was busy, with the clink of cutlery on plates and people filling mugs with coffee. 'Feeling fresh?' I heard someone ask, and people around him started laughing. It was jovial, even with sore heads. There were trimmers and vets from as far away as Canada and Brazil, a lot of them wearing their own company-branded clothing. That was something I had been noticing, more and more. It always makes me smile. It shows the pride people take in their businesses and in their vocation.

I like food a lot, but I love pastries. My mum comes to my house for coffee, every Sunday morning, and she knows baked goods are the price of entry. I particularly love Danish pastries. It was breakfast time, and I was in Denmark, so I had high hopes. It might sound sad, but I had been thinking about it for weeks. No, really. I made my way to where I thought the goods would be and … nothing. I widened my search area, scanning the room in an arc. Still nothing. In the end, I asked a waitress where I might find them, and she told me there were none. I realised I was probably more devastated than I should have been, so I consoled myself with coffee and toast.

My talk was scheduled to last an hour. There was only one event in each timeslot. That meant I would be in front of the whole conference. I was keen to see the auditorium and the tech I was dealing with. The venue can vary a lot. I've been to events like this in hotels, banqueting halls and farm sheds. The venue is unimportant, but the audio-visual setup is critical. I went through every scenario I could, just like I did at the hoof-trimming conference in Orlando, to avoid falling flat on my face.

When I was at school, I'd fake an illness to get out of giving talks to my English class. I managed to avoid every single one at my primary and secondary schools. That's not something I'm proud of, but it felt necessary at the time. I had a pathological fear of ridicule from my classmates. I can still feel the trace of the panic that tore through my system on those occasions. Even just being in the room, with those godawful orange plastic chairs and the scratched desks in front of me, if there was a chance someone might be watching me, my guts would twist, and my nerves would concentrate themselves in my bladder. The fear of wetting myself was constant.

I once got called to see the headmaster, in his classroom. I was seven at the time. I'm not sure what I'd done wrong, or even if I'd done anything wrong. All I remember is the headmaster, Mr Hannay, a naval reservist with a big black beard and a chunky frame wrapped in a brass-buttoned blazer. He spoke to me in front of his class; one that at the time included my older brothers, Bob and James. I can't remember how the conversation went, but I do remember leaving that cold 1980s classroom in my school shorts, scruffy white shirt and red and grey striped tie, having left something behind: a puddle of pee. The apprehension of just being in that room, in front of Mr Hannay, had overcome me in the worst possible way, and I had pissed myself. Even worse, it had all happened in front of Primary Six *and* Primary Seven, the upper classes in the school.

I have given talks to rooms full of farmers, students and zoologists. It's not unknown for a thousand people to be in attendance, but no matter the gig, the nerves are still there. I suppose there's an element of imposter syndrome. I can see why people might get annoyed when I put myself out there on the internet. It might seem as though I think I'm an authority on hoof-trimming theory and practice, like I'm holding myself up as a leader, but that would be missing the point. I'm an average hoof trimmer. I just happen to be the one bringing it to the wider world.

That October morning, the auditorium seemed perfect. I ticked that off my list of worries. Next on my checklist was my introduction. I had been to enough conferences and seen enough presentations to know that if you don't grab your audience here, you've lost them. I wanted to make them sit up and pay attention. The best way I could do that was with something they didn't expect. I needed to use some misdirection, sleight of mind, if you like. I had something that might work. At least I had thought it might, when I was putting it together at home, in the safe space of my office, at 3 a.m. when all things seem doable. But there, under the burning stage lights, in front of four or five hundred of my peers, people who were proud to work in agriculture, I realised I might have overdone it. I might be about to push them too far.

The house lights dimmed. A brooding, suspenseful melody filled the speakers around me, rising in pitch as the screen lit up behind me. On the screen, a drone flew through a darkened, monochrome dairy facility. Cows moved slowly below. My voice faded in, deep and echoing, and I prayed my bluff would pay off.

'Cows are supposed to live outside on the green grass of lush pastures, but instead they're subjected to a life of misery, in huge concrete sheds with little fresh air or natural daylight.'

Dark images filled the screen. Cows, walking through a shed. Calves being bottle-fed in the gloom. A lone cow limping into a hoof-trimming crush. I could feel the mood of the room shifting.

'Cattle are forcibly milked twice a day by machines which have no business being near a cow. The older baby calves are forcibly removed from their mothers and held captive in small pens, and subjected to being restrained in metal crates … cows become lame …'

I willed the video to speed up. What were the crowd thinking? If I'd been nervous before, I was shitting it now.

Bang. The video stopped. The auditorium was dark. Everyone was quiet. The lights came back up. The screen was alive again. The video rewound, with increasing speed, my voice and the music squealing in reverse. Then, the video played again, but brighter, with more colour. Bouncy music played in the background, over birdsong. My voice came in again.

'When it's possible, these beautiful cows graze outside and on the green grass which is grown specifically for their needs and when the weather turns wet and cold outside these gorgeous animals are brought in to specifically designed facilities which cater for their every need.'

I could feel the tension ebbing from the room.

'Cows love routine. Twice a day, every day, they wander at their own pace towards the milking parlour, where highly trained staff capture a nutritionally incredible product to help feed our nation ... To begin with, calves are housed in smaller pens, so that their individual needs can be catered for on a case-by-case basis before they move into their larger groups, where they can enjoy the social aspect of being an adolescent bovine ...'

My uplifting cut faded out. I looked up and out, to the back of the auditorium and the higher seats, where Kevin, Craig and Ashley were sitting, and as the applause kicked in, I breathed. Buzzing now, I had their attention. I could begin my talk.

I explained the burden of responsibility I feel; that my work can be framed in different ways, that certain situations, difficult trims, lame cows, can look negative and could pull down the agriculture industry, but if I show context and give all the facts, I can let the public in. I can give an insight into what's really going on. Some of the feet I show are upsetting to viewers, but talking through the bigger picture, explaining how a foot became a certain way, helps people to understand that the farmers I work with love and care for their animals, and that they'll do anything they can to make them as comfortable as possible.

I spoke about my journey through social media, the impact it had had on the farming community. I talked about influencing the development of the industry we work in. This was the easy part. I was in a good flow, now. I love what I do, and I love talking about it. My hour passed quickly, and I ran out of time for all the questions at the end. But one question stood out, because I couldn't come up with an answer.

A man called Alan stood at the back of the room. He was tall, shaven headed, solid looking. I could tell he was a worker just by looking at him, and like his whole team, he wore immaculately branded work gear. 'With respect to the trimming side of your business,' he said, 'how do you intend to gain longevity and a business your children can eventually inherit?'

I must have given him a blank stare for a few seconds. I didn't have a good answer for him then, but his question ignited something in me. What *was* I going to leave my children? Would they inherit a hoof trimming business? Or would I be leaving something else entirely.

I told Alan that the Hoof GP Limited hoof trimming business was in my name. I built it in my name and the value of it hinges on my relationship with my customers. My kids can never inherit that completely. They would need to build their own reputations and build trust with their own customers. But would they even want that?

Hoof trimming is marmite. You love it or you hate it.

You're covered in 'marmite' from head to toe, hour upon hour. It's physically demanding, mentally draining, and it's a financial powder keg. You chase unpaid bills and hope that the cashflow gods are kind to you. I'm not sure my children would want that as a career, but I have been thinking about Alan's question a lot. And since that day, I've done everything I can to build a future for my kids; something I can pass on. That's why we bought a couple of houses, and why we are going through a pretty monstrous renovation project at Carsegowan right now. Holiday cottages and rental properties don't need a face, and they don't rely on years of relationships linked to the owner of said face. That hopefully creates a more stable future for our kids, one where they have choices.

Of course, should they choose to be covered in cow shit all day, I will support them all the way, and I suppose, on the flip side, the Hoof GP brand isn't necessarily tied to a person – even though it incorporates my initials. I could be like Dr Who – or maybe Dr Whoof – regenerating, every few years. Maybe one of the kids will want to be the future Hoof GP.

Before my talk, I had been in my own wee bubble, with Ashley, Kevin and Craig. I'd already spoken to a few people I hadn't met before. Afterwards, everything changed. I had spent all that time worrying about presenting something worthy of my audience – something they might find interesting and relevant – and now relief flooded my body. A dozen people I'd never met approached me to say thank you, as I left the stage. They told me how interesting they had found my talk, and my head swelled, just a bit. I suppose we all want the validation of our peers. In my case, it made me feel that little bit more legitimate.

The rest of the conference was a blast. I met a huge cardboard cut-out of myself in the hallway: an advert for Hoof Grip Pro. I got a sneaky photo, then spent the rest of the conference avoiding 2D me. I didn't want people to think I was too proud of my doppelgänger, but I was (and still am) proud of the product, and the sight of the ad threw me, and made me smile every time I passed. I kept thinking, *How did I get here?*

Dinner was a highlight. The chance to sink a few beers – or fruit ciders, in my case – and cut loose with some like-minded individuals was not to be missed, and having the whole team there only added to the air of celebration. The highlight was a charity auction. It was a good opportunity to give something back and have a lot of fun doing it. I ended up bidding a thousand euros for a Japanese hoof trimming knife that now lives under a pile of clothes in my spare room.

Craig fully embraced the festivities, sinking beers and shots with abandon. I challenged him to bid for a hoodie and promised to pay whatever it cost. He didn't look too sure. Craig and I like to play the odd practical joke on each other, but he started to bid, wondering if he was about to get stuck with a bill. 'Forty euros. Sixty euros. Eighty. A hundred and twenty.' The number kept going up. Craig looked more nervous, now. He looked across at me for reassurance.

'You'd better fucking pay this,' he said.

'You'd better fucking win this,' I replied.

The auctioneer called it at two hundred and seventy euros. I told Craig he could pay me back in instalments. He shook his head, but I knew he had enjoyed his moment in the spotlight. Then he proved it by climbing onto the stage, and joining in. I filmed him on my phone, stalking back and forth, like a miniature tiger, with a set of cowboy boots on his hands, urging the bidders higher.

A Dutch hoof trimmer took a bit too much of an interest in him, and Craig, being Craig, didn't want to upset the man. We had donated some Hoof GP merch. One of our hoodies was next under the hammer but the Dutchman was having none of it. He wanted one with a side helping of 'odour de Craig'. And so, three sheets to the wind, Craig began to strip. He made a meal of it, dancing to an imaginary tune, flexing, tattooed guns blazing. He peeled his sweaty top over his head with as much fanfare as possible and extracted a winning bid of three hundred euros. Then, he climbed off the stage to deal with a disappointed Dutchman who had been in the toilet and missed the whole thing.

We returned home from Legoland tired but refreshed. Now though, I had a more daunting audience to contend with: the general public. Wigtown is Scotland's national book town. Books outnumber people two hundred and fifty to one here. Each year, the town plays host to the Wigtown Book Festival, and in 2024 we held an event to launch my first book, *Bruised Sole*. *Bruised Sole* is my autobiography. It charts a twenty-five-year battle with undiagnosed Rapid Cycling Bipolar Disorder. As the titles suggests, it's raw. I wanted

Bruised Sole to be a success and to help other people going through struggles with their own mental health. I needed to publicise the launch of the book, and the best way to draw attention was with a live event. Did that mean I wanted to do it? Hell no. I was about to discuss my innermost feelings with over four hundred people in a marquee – a good few of them friends, customers and people I'd grown up with – and a whole load of Hoof GP viewers, online. I probably don't need to tell you what that meant for my bladder.

I wanted to say no. Of course I did. But no wasn't an option. I saw a challenge, and I had to complete it. That didn't mean I wasn't shitting myself. On the plus side, I was at home, in familiar surroundings, and I wasn't going it alone. My brother, Bob, arrived at the house and we lifted some weights together. We pushed each other on, taking small breaks between sets, talking shit and catching up. When I was suitably worn out, I showered and dressed. I never quite know what to wear. Clothes are a way of fitting in, a shorthand for the world, telling them who and what you believe you are. But where was I trying to fit in? I decided to go comfortable, with jeans and my favourite pair of trainers, thinking I might as well be comfortable when I messed it all up in front of several hundred friends and acquaintances. We stood in the dark of my dilapidated shed, and I downed a couple of Disaronno and Cokes. Then, Bob drove us to the gig.

Getting on that stage, mic'd up in front of a home crowd, was one of the hardest things I've ever done. *Bruised Sole* took me years to write. Then, there were months of indecision and not knowing if I should actually release the book. There are a lot of family stories in there, deeply personal stuff. Finn McCreath, a local farmer, was interviewing us and knew how to put us at ease. Finn walked onto the stage, gathered his thoughts, then announced me and Bob, so we walked to our own seats to a round of applause. As I turned to see the crowd in front of me, I knew it would be okay. It was better than okay. It was fantastic, but it went past in a blur. Luckily, I didn't fully appreciate who was in the crowd. Almost all my customers had turned up. I would never have been able to get up there if I'd known that. These are people I massively respect and the fear of letting them down would have been too much. I couldn't have hoped for a warmer reception, and the lights and full house meant that I looked out onto a sea of blurred faces. It took weeks to work out the scale, as more and more people I met told me they had been there. Thank God my eyesight isn't what it used to be.

October in the Shire also means Halloween and guising. 'What's guising?' you ask. Well, as with a lot of things, the Scots invented Halloween ... sort of. Okay, maybe the Irish had a bit to do with it too. I know in the US Halloween is a massive thing, with trick-or-treating being the main event. Here though, we do things a bit differently. In the early Celtic world, it was the pagan festival of Samhain, the time when the veil between our own world and the spirit realm was said to be at its thinnest. There's a bit of debate about it, but one theory is that the church superseded the tradition with All Hallow's Eve.

Some traditions still persist from those early days, in-grained in our modern culture. I can remember dooking – or ducking – for apples at school. I would lean over a basin of water with bobbing apples, balancing on the back of a school chair and gripping a fork between my teeth. When I had taken aim, I would drop the cutlery into the basin. The goal was to skewer one of those beauties. Other times, the rules of the game were different, more traditional and less health and safety conscious. At home we'd stick our faces in the basin with our hands behind our backs. It's a tradition some say we appropriated from the Romans. Apples were in season and symbolised fertility. Who knows what the truth of it is? All I know is we were happy, faces in the ice-cold basin, doing our best to get hold of one of Braeburn's finest.

Guising is just as it sounds – dressing up, disguising your-self against evil spirits, as you cut about, performing for treats. Traditionally, it would have been apples, monkey nuts and treacle scones, but these days there is a lot more sugar. When we were young, costumes were a bit more ad hoc. I remember my sister, Kirsty, making a fantastic witch costume out of a bin bag and a hat from thick black sugar paper. She had an old brush handle with some twigs attached, if I remember correctly. We'd walk round our local village, Port William, or Mum would drive us between farms, where friends and neighbours would be waiting. It always seemed to be blowing a gale, but we'd insist on going, carrying our supermarket plastic bags. We didn't have the pumpkin shaped buckets and deep wizard hats my kids do now. We had to make do with whatever we could find in the cupboards at home, and I think we were happier for it. We'd knock on all the doors we recognised, and some we didn't. Daring each other all the way. We never went trick-or-treating. That's slowly creeping in now, but it just wasn't a thing back then. We all had to have a party piece; a poem or a joke or something silly to say. I don't recall any of the jokes, but I know they were terrible.

I love dressing the kids up now, but it feels like they're growing too fast. It's as though they're old before their time. Keir is eleven now and he's not bothered about guising. Campbell is still fully in the mix, though. We have to watch him. Last year he went as the Hoof GP, complete with the blood-covered grinder. That's not a totally accurate representation. At least, I hope not. Campbell and his friend went as Gru and Bob the Minion the year before. They were quite a sight, wandering round the town.

Every year, the residents of Wigtown do themselves proud. Loads of people get involved, and houses look more and more like what I imagine Halloween to be in America, with pumpkins and lanterns lighting up windows, and carefully curated experiences inside and out. You might walk into a house to see someone performing an operation on a skeleton on their kitchen table or pass a witch brewing something, in a small shed on Wigtown High Street. It's exactly the right amount of weirdness to entertain you as the days get shorter and colder, and November sits just around the corner.

Chapter Twelve:
Idiots Assemble

'U got a boat?'

That's how it started.

I was in my office, editing a video, when my phone let off the usual pinging sound. Another message. I get a lot of messages these days, but this one was different. It was Cammy Wilson, with his usual style of cryptic but somehow pointed questioning. I'd say he should be a detective, or a TV presenter, if he hadn't already served his time as both. I replied as fast as my thumbs let me. I wasn't surprised by the question, but I was intrigued by the reason behind it.

'Got the world's fastest jet ski ... Any use?'

Another ping. 'Of course you do! I'm planning to go and rescue the sheep that's stranded!'

And with these three short messages, my head began to hum. Before long, a plan formed; a plan that would end in a story covered by every major news network on the planet – a ridiculous tale that would capture the hearts of millions and pop up in every tabloid newspaper from Tipperary to Timbuktu.

This short exchange led to us rescuing Fiona: 'the world's loneliest sheep'.

Cameron Wilson is best known to the world as 'Cammy from *The Sheep Game*'. For the uninitiated, *The Sheep Game* is Cammy's vlog. It follows the peaks and troughs of his life trying to stop field lice dying. It's a thankless task. They love finding new ways to off themselves. In one memorable snapshot of a post, Cammy can be seen finding one member of his flock hanging from a tree. Cammy's a former Glasgow detective trying to make it as a farmer. Most farmers are born into it, and the money involved in running a farm is only increasing. First-generation farming is tough, and rare.

A few years ago, I saw his videos. I noticed how hard he was working but I could tell he wasn't using all that effort to best effect, so I messaged him and told him. We've worked together on a lot of things, so I've observed him up close. He's ridiculously tall, with blond hair curlier than mine, and I think he might win the prize for most energetic grown-up I know. He's like a six-foot-three springer spaniel: lanky, floppy-eared and all over the shop.

It was late October 2023 and a story about a sheep stranded on a remote Scottish beach was snowballing online. Two years previously, a passing kayaker had spotted the sheep running along a beach, at the foot of some imposing cliffs on the shores of the Moray Firth. She hadn't thought much about it at the time, and she carried on paddling through the cold of the North Sea.

A couple of weeks prior to my message from Cammy, the kayaker, Jill Turner, followed that same stretch of coast round the top of the Black Isle, and saw that same sheep strutting her stuff up and down the rocky beach, bleating for attention. The sheep had obviously been there all along, and she seemed to be stranded. Jill pulled out her phone and took a picture. Then she contacted the *Northern Times*, part of the Times News Group, and from there the story grew arms and legs. The loneliest sheep in the world began to capture people's hearts. Then the wider media stepped in, then the SSPCA – the Scottish Society for the Prevention of Cruelty to Animals – and then Animal Rising, a group of vegan activists. Soon, different groups were spinning the narrative to suit their own purposes.

'The farmer and his family are getting death threats,' Cammy said. 'We should help them if we can.'

I pondered Cammy's request for an hour. I was frustrated that I didn't have the perfect craft for this situation. I know that's ridiculous, but for some reason I thought I should. I did some googling, and after careful analysis – i.e. time spent down a digital rabbit hole – I concluded that I should buy a boat. I fired a message back to Cammy with a photograph of a RHIB, a rigid-hulled inflatable boat, that might do the trick.

My phone pinged almost instantly. 'Calm down,' the message said, with the obligatory laughing face emojis.

Ashley sat opposite me on the couch. She had heard my phone, so I showed her the picture. 'There's no way you're getting a boat,' she said.

The voice of reason had spoken. I sent Cammy a picture of a superyacht.

I felt giddy. Like a small child. The thought of it; the idea of buying a toy, or a trip to a theme park, the anticipation is almost always more exciting than the act for me. I didn't expect any of it to happen. It was a flight of fancy. Madcap schemes danced in the consequence-free environment of my mind's eye.

Fiona, as she would soon become known, was stuck at the bottom of a three-hundred-foot cliff in a part of the Northeast of Scotland known for its rugged beauty. That says something in a *country* well-known for its rugged beauty. I got more details from Cammy. She'd probably fallen down there as a lamb. The farmer had been trying to rescue her for a while. The SSPCA and Fire and Rescue had taken a look and decided it was too dangerous. Now, with the media focus on the scene, Animal Rising activists had set up camp on the farm. They had started feeding the sheep on hay, a type of dried grass. She was surrounded by *fresh* grass, so it was an interesting strategy, one that exposed a fundamental lack of understanding, when it came to sheep.

I grew up on a farm. My dad was a farmer. I work with farmers. As a group, they tend to have a 'let's go' approach to most things. If a farmer hadn't been able to get hold of one of his own sheep, I didn't hold out much hope for a bunch of protesters. I have nothing against vegans. I like them, though I couldn't eat a whole one. I respect their sentiments. They care for animals, care about their lives and how they are treated. I care about those things. So do the farmers I work with. I like to think if I was an activist I would have gone about things a bit differently; but they were just getting started.

There was an online petition, trying to force the farmer to do something about the sheep. It was at forty thousand signatures and climbing. Messages flooded Cammy's inboxes, from fans, asking if he could intervene. If we were going to do this, we needed to do it soon. We talked through the options by text. We were thinking of something that might work in the next week or two, but we didn't have that much time. We bounced ideas in between the usual chat about how many hoodies we should order. The Hoof GP LTD sells clothing and merchandise to a global audience. Cammy and his team help make that happen, dispatching orders to the world from a warehouse in Ayrshire. We talk a lot, and this was a welcome distraction from logistics.

The following morning, Cammy was back on the phone. Activists were putting more pressure on the farmer, he said. The threats were increasing. This was no longer a pipe dream. We needed to get moving. So, I called my big brother, James.

'What do you want, you little £@%^?' James said.

'Fancy a weekend away?'

I pitched it to him as a challenge no one else had managed to complete, knowing he wouldn't be able to resist. He once nearly became a professional golfer but decided it didn't pay well enough. He's a little bit competitive, and quite possibly the most single-minded man I know. Middle child syndrome I say, even though technically the middle child in my family is my sister, Kirsty. Cammy recruited Als Couzens, a mutual friend of ours. Als is an artist and designer, a bit of a genius. He loves painting sheep and cows, so this was kind of a good fit. Cammy also recruited Ally Williamson, who he described as 'a mountain goat'. I'd never met Ally, but I knew he was a sheep farmer from the Isle of Lewis, so probably more experienced in getting sheep out of tight spaces than we were. He was younger and fitter than all of us, the border collie to Cammy's spaniel, so handy to have on your side.

Cammy and I went over our rough plan one more time. We agreed that whatever happened we were bringing the sheep back up. We both knew we'd make it work if she was one of our own. I loaded my Polaris off-road buggy onto one of our trailers. As vehicles go, it's robust. I had driven it up rock faces and through bogs. I once jumped a pick-up truck using some makeshift earth ramps. It's a beast, even if it isn't much use for the school run. I mean, I *have* used it for the school run, purely for the entertainment of my kids, but I never thought it would have a practical application.

I hooked the trailer onto the back of my pick-up and headed off. I picked Cammy up an hour and half away, in Ayrshire. Then I traversed the country and grabbed Als and James in Perth. Als is in Edinburgh and James lives in Fife now, but we won't hold that against him. The atmosphere in the pick-up was boisterous, with laughing, joking, and a whole load of innuendo. It felt like a lads' day out. We stumbled into Tiso, a well-known outdoor store in Scotland. There was everything we could have needed, for any outdoor shenanigans we could think of. And there were a lot of people who looked like they knew what they were doing. We hovered round a vast selection of climbing-ropes and argued for a while. Cammy reckoned we only needed one length, probably because they were £180 each. I said we needed two, and he couldn't quite bring himself to admit I was right.

Eventually, the shop staff got bored of watching our debate from a distance and waded on in. An older lady approached first, clearly aware we were up to something daft. I remember her laughing; she was the sort of person I couldn't help warming to, like I'd known her for a long time. 'Hey, you know the sheep? These guys are gonna rescue it,' she shouted to her younger colleague. Cammy held their attention as he explained our masterplan in detail. They'd been following the story online and seemed as engaged as we were.

I won the argument, and we got back on the road, with two lengths of rope. We travelled the A9, cutting through achingly beautiful landscapes, through the heart of the country and the trees and hills of Perthshire and on to the craggy-sided glens of the Cairngorm National Park. I sometimes think I need to get out more. I live in Galloway, on the shores of Wigtown Bay and the edge of the Southern Uplands. It's like Scotland in miniature, with the sea on one side, moorland in the middle and mountains to the north. I love home, but the Highlands are another level. You're dwarfed by your surroundings, in a place where two continents collided a hundred and fifty million years ago and you can still see the join.

We arrived at the farm in the dark, after a seven-hour journey. It was early November, and the nights were longer now. It was peaceful. Angus, the farmer, and his family, welcomed us into the warmth of their home. If there's anything more inviting than a farmhouse on a night like that, I've yet to see it. We passed through the flag-stoned hallway, past muddy boots and hanging coats, and I couldn't help being aware that it was late, and we must be imposing, like we'd just arrived home from the pub, ready to keep the party going, and the neighbours were trying to sleep. We hadn't had a drop, but that was about to change.

Angus had a well-to-do way about him. He introduced us to his wife and two daughters, who were Hoof GP fans. After the introductions, we got down to business, fleshing out our plans. We studied drone footage of the cliffs and the shore-line. Angus told us about the hassle he had been getting from different groups of activists. It was clear from the footage that there were worse places to be stranded. There was plenty of fresh run-off water from the land above the shore. There was green grass, and there were caves for shelter. I knew people who would pay to stay there. The sheep might have been living the dream, if she hadn't been a sheep. They're social animals and she seemed to be approaching kayakers because she was lonely. Normally, a sheep who hasn't seen humans for a while will run. Lonely or not, we needed to get her away from there and take the heat off the farmer and his young family.

Angus's wife cooked us a rich, juicy beef stroganoff and supplied us with beer and red wine. Nine of us sat round a massive wooden table. The boys and I were in a stranger's home, and it could have been awkward, stilted, but the mood was relaxed. We chatted about how strange it was, finding ourselves in this situation. Everyone else was on beer so I had to finish the entire bottle of red. It would have been rude not to.

The farm sat on meandering cliffs, near the entrance to the Cromarty Firth, overlooking the North Sea. The family rented out pod-style holiday homes to visitors for extra income. One of these was ours for the night. When we got there, I could see it had been made from rounded wood, like two giant barrels, joined to form a cross-shaped plan. I felt like I was in the opening scenes of *Lord of the Rings*. The wine had gone to my head, and I was tempted to jump in the hot tub, but we needed to be up early, so Als took a room, Cammy slept on the couch, and James and I bunked together in a twin room, like when we were kids, minus some of the arguing and the mess. It was a comfortable throwback to childhood and planning the next day's adventures. We talked for a while, nestled in fluffy Egyptian cotton, and reminisced about old times. As I drifted off, I found myself saying 'goodnight', in a way we probably never did as kids.

We were up early to see an amber east coast sunrise emerging from a cloud inversion. We stared out on clouds, blanketing the dark sea below. It was otherworldly, disconcerting. It might be the most beautiful sunrise I've ever seen. I hoped it wouldn't be my last. Our window of opportunity was short. Angus had come up with a ruse to clear the way. He had told the Animal Rising activists that he would be moving cattle in the fields around the cliffs, near their access point. He suggested it might be a good opportunity for them to go and replenish their supplies. He told them they could return mid-afternoon, once he'd finished moving cows. We needed to be quick. The last thing we wanted was a confrontation. They had been there for a week, planning whatever they had in mind. Any encounter risked conflict and inflaming the whole situation.

Ally the mountain goat arrived to join us. He looked sturdy – which I was glad about – with an impressive mullet, and curls that gave Cammy and me a run for our money. Anyone rocking that kind of do couldn't be short on confidence. Ally speaks Gaelic. English is his second language, so he has been taught to speak it properly, unlike the rest of us. The five of us made our way down the farm road and across the fields that lead to the cliffs, along with Angus, and some additional helpers, in pick-ups. Cammy and I drove down in my buggy. On the way he mentioned something about wanting to stay alive, and I was inclined to agree. We both laughed a bit too much.

Until that point, all we had was drone footage and aerial views, courtesy of Google Earth. It had given us a rough understanding of the terrain, but not much more. We crossed perfect green fields, to a barbed wire and mesh boundary fence held in place by ageing wooden posts. We leaned on the fence and peered over the edge. Now we had more of an idea of what lay ahead, and what might be possible. I realised what we were seeing wasn't all cliff. We were very high – the cloud was still below us – and the way ahead was steep. There was no way we were walking down there, but to say we had to climb down cliffs was – thankfully – a bit of an exaggeration.

Our Animal Rising friends had already strung a rope through a gap between two outcrops, leading down to the shore. They had anchored it on a tree that had seen better days, but we couldn't argue with their results. They had tested it, after all, leaving a path where they had beaten the grass, as they made their way up and down.

We agreed that Cammy, Ally and I should clamber our way down to the shore and try to capture the sheep. James and Als would stay with the buggy for stage two.

The sea haar was beginning to lift. Time was tight, so we began our descent. Cammy started out in front. I could see the tension in his frame as he edged forward. The mist had left a heavy dew on the grass. It was wiry and slippery. We descended, one footstep at a time, in a semi-controlled fall, hunting for purchase with our heels. Again and again, I lost my grip. I'd grab the rope to get my footing back, then repeat the process for a few steps. I wasn't alone. We kept going, falling and catching ourselves, slipping and sliding, regaining composure, then starting again. Through all of this, Ally strode ahead. I could see now why Cammy called him the mountain goat.

A few minutes passed before we reached a sheer rock face, and a twenty-foot drop. If I was a proper outdoorsy type, I'd call it 'technical'. We agreed our only option was to mock-abseil down. The fog was lifting but the surface was damp. Water seeped from the rock. I took my time, threading the rope through my fists, trying to avoid friction burns.

'Don't put too much trust in the rope,' Cammy said. His

voice was higher and tighter than normal. I could hear pieces of rock tumbling above us, then the dull thud as they hit grass below. I had no choice but to trust the rope and the looped handholds tied at three-feet intervals. But we knew it was a gamble, and we were all smiles when we reached the bottom.

I kept thinking of The Knock, the farm after Barmeal, where I grew up with my brothers and sisters. The Knock sits up on a similar shoreline, but it's over on the west coast. The view is at its best there at the other end of the day, when the sun dips into the Irish Sea, turning it orange, then red, then purple. We spent a lot of time exploring that coastline, as kids. We'd traipse over grey, sun-baked pebbles, to a place we called the back bay. We'd feast on sand-filled sandwiches and hot juice. We used to think you could fry an egg on those rocks. I think we might have tried it. Those are some of my happiest memories, but there was always a sense of danger, and a knowledge that the rocks and the sea would end you if you let them. I remember Dad having to rescue one of his precious Charolais calves from the rocks down there. In the end he had to call out the coastguard.

We had made it through the vertical part of the descent. We could see a cove, below us, littered with debris from old rockfalls. The ground was steep and loose, with stones like fist-sized marbles, carpeted in ivy. It looked a lot like tailings, or waste rubble, as though the cliffs had been dug out. We took it in turns to slide and skate further down a two-hundred-foot bank. When we weren't slipping, we were tripping. The bay emerged in front of us; the place the sheep had called home for the last few years. It was a good five acres, a lot of it green, nutrient dense grass. The shoreline was craggy, jagged and bleak. Waves crashed onto the beach then dissipated through small channels between black rocks. This was prime shipwreck territory. The boat would have been a bad idea.

Angus had told us we would find the sheep cozied up in a group of caves, on the far right of the shoreline. We stumbled our way in that direction, following a path she had worn as she made her way back and forth to higher ground, and all that grass. We couldn't see a cave at first, but our view changed as we moved round and down, and slowly, a rocky mouth rose up to meet us.

We reached the entrance, and I went in first, treading lightly over small pebbles, scraps of driftwood and seaweed on the stone floor. It had that salty, musty smell all seaside caves do. I could see the seams where the rocks had fused together with heat and pressure. And I could see a sheep, standing at the back, in a shaft of light beyond the darkness, as though she'd been waiting. It was surreal, like a dream-scape. The cave was part of a network, with three exits; a giant rabbit warren carved from the rock by time and tide. That explained the backlighting. This was going to be tricky. Whatever move we made needed to be controlled, and fast. We edged closer, in a row, as one. Sheep don't like sudden movements, and this one hadn't seen many people in most of her lifetime. This was her home, and she had no idea what we were about to do.

Ally broke off first, closely followed by Cammy, then me. A big part of our plan was the need to get her in the cave. A flock of sheep is easy to catch. They move as a congealed mass. One sheep is a nightmare. She turned and ran, and my heart jumped up a gear. If she got to the other exit, we were going to have real problems. This was her domain. Sheep are much faster on this kind of land than we could ever be. She had four legs, and she was happy climbing. We couldn't even walk in a straight line. If we didn't get her now, we'd be here for hours. Her fleece was overgrown and if she bolted for the sea that mass of wool would drag her down. She'd weigh so much, the three of us could never pull her out. She would drown.

Ally darted after her. She turned and ran for another part of the cave. Ahead was what looked like a tunnel. Cammy moved in next. I crept in closer. I didn't want to leave any exit clear. I didn't want to be the one to slip up. I heard movement inside. Ally had made a grab for her. I moved further into the gloom, spreading myself wide across the mouth of the cave. But as my eyes adjusted to the darkness, I could see I was surplus to requirements. The sheep had run into a dead end, and the warm embrace of a couple of sweaty farmers.

She struggled at first, but slowly, she settled. We could see her properly now. She was in good condition, and very well fed. Despite her overgrown fleece, I could see she was a powerful animal. She could have made things a lot more difficult for us. We made our way back out of the cave, Ally and Cammy walking her, gripping her wool with weathered hands. We reached daylight, then traced our steps back out across the rocks and up to the grass and the rope. Now the real work began.

We walked her the first fifty or sixty yards, and she fought us all the way. Every few steps she dug her legs in, hard. Sheep's legs might look spindly, but they're strong, and she had four of them. I could feel the strain in my back and shoulders and the muscles of my own legs. The sun was higher now, and the fog had gone. It had been useful insulation, from what now felt like unseasonal heat.

We tipped the sheep onto her back, using her fleece as a mattress. That would hopefully soften the impact of any stray rocks. Cammy and I grabbed a front leg each and Ally took her back legs. Her head faced up the hill, so Ally had his work cut out. We counted 'one ... two ... heave!' and lifted with everything we had. That might sound like a brutal way of moving her, but it was the safest option. This is how farmers have sheared sheep for generations. On her back, she couldn't run away. That would prove fatal at just the wrong moment. We pulled her up the hill, each of us stripping off layers of clothing as we went. I abandoned a perfectly good Hoof GP hoodie. This was a tough shift. We carried on, pulling and pushing and lifting, heaving her upwards, until I lost track of time. I had no idea how long we'd been at it. I felt physically sick, with the exertion of each lift, and angry that I'd become so unfit. The ropes we had bought in Tiso weren't quite long enough to reach the shore, but we had run them down as far as they would go. We were so very nearly there. One. More. Lift.

Fiona was calm, now. There she was, being manhandled by three rough men she didn't know, getting dragged up a hillside against her will, and she just lay there. She couldn't know that we were trying to help her, but it seemed as though she did. I felt the oils from her fleece on my hands, and the soft fur on her legs between my fingers. We reached the end of the rope, and our secret weapon, a Tarff Valley feed bag – a large red and white woven tote bag, with lifting straps, designed for a telehandler. Normally this was packaging for animal food, but today it was a makeshift hammock, delivering a more precious cargo. Yorkie had given it to me when I came up with this part of the plan and suggested he might like some free advertising. He couldn't resist the chance to see his branding in the rescue.

I slumped on the grass, holding onto Fiona while Ally climbed up to retrieve the bag and attach it to the rope, and Cammy phoned James. We had struggled to stay on our feet coming down. Now I was trying to hold on to a two-hundred-pound sheep, by myself, and I realised she was slipping. Worse, she could feel the downward movement. Her legs flailed as she twisted her body, trying to get the momentum to roll back on to her feet. Sheep are good on their feet but lousy off them. They're so bad at getting up, they can die just by getting stuck on their backs. But Fiona's inability to right herself didn't make me feel any easier as she slipped again. I tried stroking the soft fur on her face. She flapped around and slid down another foot. I let out a yelp, knowing we were in trouble.

'You two want some alone time?' Cammy said.

I looked up and saw him walking back towards us. I wanted to hit him with a witty comeback, but I was relieved he was there. I didn't fancy being the man who let the loneliest sheep in the world roll down a hill into the North Sea as millions looked on. With three of us here, she was a lot safer.

We opened Yorkie's bag, spread it wide on the grass and rolled Fiona inside. I had time to look around for a moment and take in the scene. Here we were on a remote beach, under an autumn sun, with a sheep in a bag. It felt good to have got this far, but there was still work to do. The bag was attached to our much-debated rope setup. High above our heads, at the other end of the rope was my buggy, a winch, and James and Als. James was manning the winch. He's someone you want on your side in a situation like this. I mean, there *aren't* really situations like this, but when the pressure is on, he's someone you want around. His RAF years have left him with a commanding presence. Ordinarily I'd worry with someone else in charge. Maybe I'm a control freak, but I knew Fiona was in safe hands with James pulling her up the hill.

We had a problem, though. The winch was useless. It will drag a buggy out of a muddy hole. It's great for dragging other buggies out of muddy holes – I knew this because I had tested it a few times, just to make sure – but here, it was just too slow. And slow isn't any good when you're trying to keep animals calm. James's solution was simple, but ambitious. He would tie the rope to the front of the buggy, then reverse, as slowly as he could, for about twenty yards. Then he would stop. The volunteers at the top of the hill would untie the rope and hold it in position. James would drive forward and they would reattach the rope and do it all again. It would work like a giant, motorised ratchet. That was the theory, but the ground was wet and heavy with dew. The tires slipped and spun as he reversed, grasping for traction.

For three quarters of an hour, we climbed back up the hill, and all I could think about was how exhausted I felt. Fiona was almost hidden from view, tucked up in her red and white sling. Ally had stopped and cut a hole in the bag using a stone, because we had got all the way down to the beach and realised we had forgotten to bring a knife. He had positioned the hole so Fiona's nose poked out of it, and she could breathe the fresh, sea air and stay as calm as possible. But as we pulled with everything we had, trying to get her back to safety, we could see her jaw moving. She was eating, ripping grass out of the ground as we moved along. We were working our asses off, and she was snacking.

We took it in turns to support her and guide the bag upwards. The rope stretched and slackened, jolting tight, whenever James reversed the buggy. We were at the steep drop again, when I realised it was the most dangerous part of our plan. We had already abseiled down the way. Now we were trying to send a sheep straight up, in a giant, agricultural shopping bag. One slip and she was gone.

We moved her to the foot of the drop, to avoid any collisions. The rope snapped back again, and we scrambled to guide it. The buggy reversed out of sight, above us. Fiona dangled in mid-air. If the rope gave way now, if a knot went, if it caught on a rock, she would break a leg or roll down to the sea. My head thumped with the pressure. Then, in one smooth motion, the bag and Fiona sailed up to the hillside above. From here it was a straight climb over long, soft grass and brown, wilting bracken, and a gentle cruise to the end. We were exhausted. She was moving so fast now we couldn't catch her, but it didn't matter.

We reached the old fence at the top. James, Als and the rest of the team were waiting. With one last almighty heave, we lifted Fiona clear over the top of the barbed wire and set her down in the field on the other side. Finally, she had been rescued from paradise. I wasn't sure she'd thank us.

We undid Fiona's bag, unveiling her to the world. Dougie, an inspector from the SSPCA, took his time to look her over. When he was satisfied, he declared her 'very healthy', but noted that she was 'overfat' from all that grass. He gave her a condition score of four and we posed for the obligatory photos. We had no idea just how far these would go. We loaded Fiona into the back of my truck, and she settled down nicely. She must have been ready for a rest. It wasn't even lunchtime, and I was done in. Failure hadn't felt like an option. I've never really believed anything is impossible, but sometimes, when you're up against it, doubt has a way of finding the cracks and seeping in.

Our next challenge was security. The interest of the media and the activists and everyone else meant that we couldn't just do what you normally would and return her to the flock.

That would still leave Angus and his family with a problem. We needed to come up with a good ending to the story, one that worked for Fiona. She is a sheep, and as a species, they tend to look alike, so the obvious thing was to make her disappear into another flock and, ultimately, into obscurity. Ally or Cammy could have done that quickly, with no fuss. But the possibility of more uproar and threats directed towards whoever took her in was too much of a risk. We needed to put her somewhere safe; somewhere everyone could be sure she would come to no harm. Cammy had a plan.

The Animal Rising crew had surely replenished their supplies by now. They were probably heading back to the farm. We needed to avoid any conflict, and any more stress for Fiona, so we loaded everything up and set off on our journey home. We were all buzzing now, caught up in the excitement of the day. We wanted to share the feeling and put everyone's minds at rest, so Cammy uploaded a short video to tell everyone that the world's loneliest sheep was no longer marooned, and that she was safe and well. I hung around in the background, either slightly hungover or suffering from a migraine. I couldn't quite decide which, but stress and excitement tend to trigger these headaches, and so does a bottle of red wine.

We explained how it all unfolded to the world of social media, and we announced that the sheep was now called Fiona. The name? In 2004, a colossal merino wether – a castrated ram – was found wandering near Bendigo Station in Otago, New Zealand. He had managed to avoid being shorn for six years and he'd grown a fleece big enough to make twenty-seven large men's suits. He was so famous in his home country, he went to meet the prime minister. He had avoided the shearers by hiding in caves, and so they dubbed him Shrek. As anyone who has seen the *Shrek* movies knows, the main character's love interest is Fiona. Fiona is a Scottish name. It was a perfect fit.

If we had been expecting a quiet, anticlimactic journey home we were wrong. The drive was busy. The world's media had seen Cammy's video, and his phone didn't stop ringing. As I drove, he sat there, fielding calls from media agencies, news outlets and production teams. One half-heard conversation stands out more than most. A producer from the UK's biggest daytime TV show, *This Morning*, wanted to know if we had a sheep in the back of my pick-up. I watched Cammy in the mirror. I could see Fiona looking at him through the rear window. 'Might have,' he said.

After the call we got the other half of the story. The producers had flown a TV presenter and vet, Dr Scott, and a camera crew to the north of Scotland. They had chartered a boat, and they were sitting in a harbour about to set sail to rescue Fiona. We couldn't stop laughing at the thought that something quite small, something we'd done as a boys' day out, the kind of thing we would do just to help each other out, had become such a big thing that a TV production company would go to all this expense. Everyone had planned big expeditions, and five dafties got it done on a whim. There were a lot of requests for interviews. Everyone wanted to be the first to get the story, and a shot of the world's most famous sheep.

We talked it over and decided the best place for Fiona was Dalscone Farm Park, in Dumfries. It's a fantastic place, run by Ben Best. We knew Ben would keep Fiona in the lap of luxury. Not that she needed too much more of that. Cammy made the arrangements, and we were all happy with what was a good solution. This way she would still be in the public eye, and everyone could see she was well looked after. For now, she had had a stressful day and the journey to Dalscone was a stretch. We dropped Als and James off back in the Central Belt – where most people live in Scotland – and headed west to Cammy's neck of the woods. It was the right decision.

Animal Rising were onto us, quickly, and they were not happy. They knew, somehow, that we were taking Fiona to Dalscone, and they felt she was being exploited. The farm was closed for the winter, so Fiona was in for months of peace and quiet, but they wanted to take her to an animal sanctuary. A group of protestors arrived outside the gates to the farm. They had placards and probably sour grapes at hanging around for a week only to be beaten by a bunch of guys who turned up that morning.

It was a tense situation for Ben. There were threats. The protesters flew drones over his farm and his family home and did everything they could to intimidate everyone there, including the animals. But Ben wasn't backing down, and neither were we. Fiona's home for the night was a sheep shed on the outskirts of Stewarton, a small village in Ayrshire where she could chill out, away from all the fuss.

I drove home in a hurry. I had a fireworks display to organise and some barbecuing to do. If I was late, Ashley was going to kill me. The next day was the fifth of November, Guy Fawkes night here in the UK. It was originally a celebration of the foiling of the Gunpowder Plot. On the fifth of November 1605, Guy Fawkes tried to blow up the Houses of Parliament in London. He was caught when one of his co-conspirators tried to warn a friend not to attend the opening of parliament. Just goes to show, you should tell your secrets to as few people as possible. That's why only Cammy and I knew where Fiona was.

These days we celebrate Bonfire Night or Guy Fawkes night by lighting bonfires, sometimes with effigies of Guy on top of them, which, now I think about it, is a bit dark. We light fireworks, eat too much sugar and generally have a good time. For me it's not about ancient history. It's about my own history; the bonfire, toffee apples, Catherine wheels, and sparklers, and the feeling that winter is on the way. It's a chance to make memories with my own kids. That year I might have gone a bit overboard. I had to go to the shop four times. Apparently, you're only allowed to buy so many explosives in one sitting. I was a bit concerned about crashing the car and the kind of blast that might cause. Having fed a group of family and friends on sausages, I got to work on the entertainment. I moved carefully, between rows of fireworks, lighting the fuses in blocks, just to keep myself safe. The effect was worth it – explosions of red and green filled the sky. Later, my neighbours told me they stood and watched from their house. They live half a mile away. None of this helped my migraine.

The next morning, I sat on the cold concrete and watched as Cammy hand-sheared Fiona for the first time in her life. Sheep are partly bred for their wool, so unlike a lot of animals they can't naturally regulate how much of a coat they have. Fiona's fleece was immense, but it was November, and electric shears would have taken too much of her wool. As it was, Cammy removed twenty pounds, or nine kilograms of the stuff. Ashley, Keir and Campbell were with me now, and I was calmer, watching the action from my own bubble. When the wool was gone Fiona was half the size. She paraded up and down, weaving in and out of her new friends. She was so calm among humans, it was hard to believe she had spent years living like Robinson Crusoe.

Later that day, Sky News interviewed us with Fiona in the background. In all my time as the Hoof GP on YouTube, with all the videos I've produced, I have never encountered anything quite like this reaction. It was a perfect media storm: a cuddly animal, a heartbreaking tale of loneliness, and a daring rescue, with a happy conclusion mildly threatened by some pantomime villains, at the end. As I said, the story was covered by every major news outlet. Think about that – one lone sheep on a remote Scottish shoreline caused all that fuss, captured hearts, and led to various convoluted rescue plans. Then, five unlikely lads turned up to the rescue for a bit of a laugh, and it was global news. It was important enough to be talked about in Australia, Japan, Russia, Canada and pretty much everywhere else.

I can't help thinking how the world has changed. For all the negativity and the dangers of smartphones and social media, this wouldn't have been possible without one kayaker taking a picture with a mobile device she just happened to have with her, then posting it to a news agency who posted it online. Technology has made the world a smaller place. For weeks, I was inundated with emails and messages of support from well-wishers, and fans who had woken up and seen the guy they follow online being interviewed on TV. Animals are a leveller. They tug at the heartstrings of kids, grandparents, and strapping lumps of men.

What did it change for us? Not a lot. It was just an incredible experience to look back on. It didn't change my life or raise my profile. It was a good laugh, spending time with friends and doing a bit of good. That's what it's all about, isn't it?

Als took all the footage we had and combined it. The end result was all sharp cuts, explosive sound and circling drone footage; an epic edit that made every one of us look just a tiny bit heroic. He called it 'Resc-eweing Fiona – Idiots Assemble' and uploaded it to Cammy's *Sheep Game* channel. He sent us all movie-style posters he'd designed. Fiona stands on that famous shoreline, looking very Hollywood. It's pure Als. I'm proud to hang mine in my hallway, and every time I see it, I smile.

We appeared on *This Morning*, live from Fiona's new pen at Dalscone. The protestors had given up and gone home, and Scott, the TV vet, had made his way back from the far Northeast. He interviewed each of us in turn, and the crew were good sports about us getting there first. It still felt like a bachelor party. That probably wasn't helped by Cammy, as he pulled his pants down to show us the underwear he'd designed. We laughed about how differently it could have gone. That's much easier to do from a safe distance, knowing we'd survived to see another Christmas.

Chapter Thirteen:
A Very Brown Christmas

I could tell you that December here in The Shire is a fantastical month. I could spin tales of snow-blanketed, postcard perfection, leaping deer and brown bunnies playing in a pillowy Narnia-esque expanse. I could describe a sleeping winter wonderland full of berries, robins and the faint echo of children singing Christmas carols. I could tell you how amazing it is, going to work with the sun glinting through icicles on branches, and a roaring log fire in the barn, toasting our backs. I could tell you all of that, but I'd be lying.

December here can be pretty bloody miserable, and it likes to sneak up on you. It's wet. It rains sixteen days out of thirty-one, which isn't quite as brutal as the months either side, but the general scene is not what you'd expect when you've been brought up on Christmas cards and movies caked in snow. The temperature sits between five and seven degrees Celsius. It rarely dips below minus five, so it is neither cold enough to keep the rain away nor warm enough to be pleasant.

By December, the cows have been inside for a few months, safe from the worst excesses of the weather. If they stayed outside, their hooves would sink in the damp pastures and leave a boggy mess. The grazing is terrible anyway, the grass being low on nutrients, at this time of year. Time and seasonal change will restore the goodness, but in winter the cows are hiding, in concrete sheds. The farmers make them as comfortable as they can, with mattresses, waterbeds and soft rubber flooring. They feed them on silage; a sort of pickled grass, cut and chopped in the summer and stored in great pits, until it's needed.

Working as a hoof trimmer is tough year-round, but December is harder. You're cold and wet. Simple movements are more complicated than they should be. You reach out for something and bash your hand on a cold piece of metal, and though your fingers are numb, it somehow hurts more. The cows aren't in their natural environment. Their feet haven't evolved to deal with the moisture on concrete floors, and they get saturated with water and urine. Their immune systems have been battling the conditions and a perfect bacterial storm for months now, and problems are coming to a head. Farmers are working their hardest to overcome weather-related problems, and just to keep things interesting, the days are shrinking at an alarming rate. Scotland is relatively far north, on a level with parts of Canada, Russia and Scandinavia. We're in the very south of Scotland, but by the time the winter solstice hits on the twenty-first of the month, we have just over seven hours of daylight to play with.

One day, a few years back, I was working in a dimly lit shed, on a dairy farm with around four hundred cows. I hadn't reached the YouTube days yet. Craig had yet to enter the game, and I was struggling. Working alone is hard. Some people love solitude. We're all wired differently, but for me, those days were a struggle. I didn't work alone because I wanted to. I just couldn't afford any help. On this particular day, I was trying to push the cows through a milking parlour – the place they go to get milked twice a day – and down a race I had created with some gates. I had forty Holstein cows to work through, and I was in a hurry. I pushed a lame, dirty cow down the race, my shoulder hard up against her hip bones, pumping my quads as hard as I could. Every step is a little victory in that kind of situation. You have to keep going. This time, though, it wasn't working.

I turned my back up against her and reached for the bars on either side of the race. I grasped a bar in each hand and pushed on her rear end. I shoved with everything I had, until her legs locked straight, and I lifted her hindquarters clean off the ground, willing her to give in before I did. I stood there, straining, weighed down by half a cow, in the dark and the filth. There was no one around to help. I felt like I could cry. It wouldn't have been a first. I can't help feeling a touch of shame, recalling these moments. I know that shouldn't be the case. We should all be better at discussing our feelings, now. We should be more aware of mental health issues, especially in the solitary world of farming. But it's a tough old industry, and the attitudes we picked up from our fathers are hard to shift.

In the here and now, in Hoof GP world, I have a team around me. I have help, and on the rare occasions I don't, I get help from farmers. Back then, I was all by myself, trying to deal with the same volume of cows I do now, trying to cope with knackered equipment and poor mental health. I needed to show the cow – and myself – that I wasn't being beaten. I couldn't give her any ground. I had to keep pushing her forward until she gave in and walked to the front of my crush. I took a deep breath, planted my feet firmly, and heaved with my arms, trying to unlock her stiff front legs. I could feel myself heating up. It was bitterly cold outside, and I was wearing fleece-lined overalls. A freezing wind whipped my neck, and I was grateful for it. I could feel sweat dripping down inside my suit, and then, there was something else – a heat spilling around my neck and down my chest. It crept around my waist and crawled down the small of my back into my boxer shorts. This cow had emptied her bladder and bowels all over me.

If ever there was a moment I didn't need a code brown, that was it. Now I *couldn't* give in. The heat and the steam and the foulness filled my nose, then my lungs. The anger rose in my gut. I pushed as hard as I could, feet on the concrete, hands gripping cold steel. Forward a bit. Repositioning my feet. Forward a bit more, and a bit more, until something gave. The cow buckled and lurched ahead. I staggered backwards, as I heard the clink of the headgate closing, locking her safely in place.

Thank God she's in there at last, I thought. *Now I can do some work.* I couldn't have cared less about my shit-filled boilersuit then. If anything, the warmth was welcome. I knew it would pass, that in another ten minutes I would be cold again and the reek and the stickiness would drive me mad, but in that moment, I had won.

Thinking back ten years, or more, I find it hard to believe the things I achieved and the obstacles I overcame just to get through each day. From the outside, these struggles might not sound particularly grand, but now I'd find them impossible without the help of the guys who work with me. Recently, four of us had forty-three cows to deal with. That's not a tough shift, even by today's standards. Craig breathed a heavy sigh when he saw the numbers, because it's all relative.

Nowadays, we trim forty to fifty cows on that farm. Craig helps get the cows in from the fields. He walks them through the parlour into the small race we carry with us as part of our KVK setup. Cameraman Graeme takes over from there. He enters each cow's number into an iPad, then ushers her forward, further down the race. At the end of the race, each cow enters the crush in turn. Craig lifts her feet, one at a time, and I come in behind him, trimming the back feet, allowing him to get going on the front. Behind us, the whole process begins again. It's a slick routine, one that has evolved with time and practice. It's fast, fluid and hassle free, because it has to be. We need to get out of there before twelve. The cows are due to be milked again at one, by which time we should be long gone.

On that godawful day, in my shit-filled overalls, I had fifty sets of feet to do before the second milking. I needed to save time, so I tried a shortcut. Rather than bring the cows through the parlour one by one, I packed in as many as I could. I shoe-horned thirty-five of them in there. They were quite happy, and they were close at hand, but there was a downside. I would trim a cow, lifting each foot in turn, updating my records against her numbers. I'd move back and forth as I needed to, searching for blocks or glue, fighting for space, and then I would move onto the next one.

The full parlour made movement tricky. It had been designed for space, to allow the cows to turn, as needed. It was cramped but it was too wide for single-file queueing. Instead of gliding between cows, as I'd imagined, I clambered and squeezed and wedged myself between them. I'd climb over one cow to try to separate the one I needed next. I gripped with my heels on the concrete, pushing and manipulating the girls in any way I could. I had years of experience then, but that didn't make it easy. They put their heads down and pushed back. They kicked and jostled and fought me. I kept on going, conscious that I needed to be out of there by midday, but knowing how easy it would be to fall to the parlour floor, to be trampled.

I moved from cow to cow and problem to problem, every time I went to that farm, fighting to get my fifty done, and I never quite got there on my own. The frustration was immense; getting kicked and headbutted, risking my neck, and losing every time. I wasn't used to any of this. Freeze branding might have been tough, but at least I had help, and a sense of camaraderie. Sometimes, that's all you need.

I remember a time on a nearby estate that still makes me laugh today. The property is home to a large dairy farm, around an hour's drive north-west from my house. I used to go there twice a year, in December and January. On this occasion, we had a hundred and twenty or so maiden heifers – heifers who haven't given birth to calves – to freeze brand.

At the time, a guy called Sam worked there. Sam was the son of one of the other workers, and like a lot of people, he may have been encouraged down that route against his will. Farming gets in your blood, but Sam hadn't inherited his father's knack for it. There had been a breakout on a neighbouring farm when I arrived. Eighty-one cows had been wintering in a field and had decided that the lure of the lush, green grass in the next pasture was just too much. They had smashed through a gate, run down the road and made themselves quite at home. The farm was owned by the same estate, so Jackie, the farm manager, gave Sam the task of rounding them up and making sure they were all safe. As Sam climbed onto the red Honda quad bike, Jackie shouted after him, 'Remember to count them!'

An hour passed as we worked away. I noticed a couple of tense remarks from Jackie, but I didn't think too much of it. I was far too busy applying my brands and taking the obligatory kicks to realise what else was going on around me. Timewise, I was only interested in the thirty seconds it took to leave a good, clear number.

Soon enough, Sam roared back up the farm road, triumphant. He had rounded the cows up on his own and returned them to their field. From the grimace on his face, I guessed he was surprised but happy at the outcome, like he'd gotten away with something.

'Did you count them, aye?' Jackie said. There was a note of caution in his voice.

'Yes, I counted forty-three,' Sam said.

'Ah fuck!' Jackie said. 'There should've been eighty-one!'

Sam's face crumpled into a frown. There was a long pause. When he spoke, he was quieter. 'I didn't count them all ...'

I'm not totally sure what was said after that. Maybe it's my poor memory, or maybe I have blocked it out, somehow. I do know Jackie wasn't happy, and that the air turned blue for a while. More often than not, when I went there, I came home with a story. There was the barking dog that annoyed Jackie and me all day. We never saw it, but it produced the kind of racket that would make anyone question their love of animals. It turned out to be a piece of tin roof, rubbing against another in the wind. There was the time one of the farm hands asked how long it would be before someone would have to go around with a white marker, to colour in the freeze brands because, 'well, how else do they turn white?'

People make Southwest Scotland a special place, and no more so than at this time of year, when the sun doesn't rise much before nine and hightails it over the horizon at the back of three. When the days are short but seem so very long because of the lack of light, you need that Scottish humour. The hardened expressions I see on the farms I visit aren't just an effect of the weather. They reflect a stoic humility in the face of adversity. Hard as I try – and believe me I do – I can never fully capture that spirit in my videos. Arguments are a form of bonding here. If we're not making fun of each other, it's either because we haven't quite clicked yet, or because we never will. Besides, what would be the point in making fun of Craigie Boy and Cameraman Graeme behind their backs? I wouldn't get to see their reactions.

For me though, family is at the absolute heart of our community, and this is very much a time of year for family. Christmas is a special time here, and not just for religious reasons. Here its roots go deeper. In the extremes of darkness and light you can feel the echo of a longer history; of Vikings celebrating Yule and pagans doing whatever the hell they did in the stones they aligned to the winter sun. It's a punctuation mark; a time to pause, turn inwards and reflect on everything we have. It's an excuse to congregate, as families, to argue, to eat and drink far too much, and to appreciate that for all that it is. I often compare Mrs HGP to Mrs Claus. She loves Christmas and dresses the entire house in festive ornaments and brightly coloured garlands. There are Christmas trees everywhere, and weird gnomes she calls gonks.

I make apricot stuffing for Christmas dinner every year. Last year, there was a bottle of Disaronno Amaretto on the kitchen counter when I started. I'm not sure why it was there, but I couldn't resist opening it and taking a drink. I gently simmered the apricots with the cranberries, reducing them enough to allow the cranberries to burst like little bubbles of flavour. After a couple more shots I added pork sausage meat, then herbs and spices, and wrapped the whole mix in streaky bacon. It's a sticky sweet but firm meatloaf and it tastes incredible. Unfortunately, that means I have to make it every year. It was only when I finished last year that I realised I had accidentally nailed the entire bottle of amaretto. I was flying, three sheets to the wind, until I collapsed in a heap on the couch.

I woke to an almighty hangover the following morning, knowing I had Christmas dinner to get through. I detest hangovers. I'll do anything I can to avoid them. I'll down pints of water before bed. I'll take Alka Seltzer. I'll wake up intermittently in the night and climb out of bed to take more painkillers, just to bypass the grinding in my guts and the ache behind my eyes. So much for a long winter's nap. That Christmas morning, I sat bolt upright, remembering I'd forgotten to bring in the boys' presents and set them in place the night before. I needn't have panicked. While I was pissed and unconscious, Ashley had done it all.

Shame-faced, I forced myself out of bed at 6 a.m., in time for the boys to shuffle, sleepy-eyed, from their own rooms. They still believed in Santa Claus at this point, or at least they told me they did. I love that notion. I want to hold on to it as long as possible for them, but I am painfully aware that they live in the real world. As Keir hits high school, others will know that Santa is a blissful myth. I don't want anyone to make fun of him, so it's a constant tug of war in my head. Don't tell him, and risk him being ridiculed, or tell him, and risk Campbell finding out the truth about old Saint Nick.

I made it through the morning, surfing the buzz of the kids opening their gifts. I almost forgot I was 'ill', until I opened a present from Derek and Nicole. I knew what I was dealing with as I tore the paper from the neck of the bottle. My stomach grumbled and my mouth began to water, and not in a good way. Of course. Another bottle of Disaronno.

The day sped past in a shimmering blur. We spent most of it at Ashley's stepfather, Wallace's place. Wallace lives in a normal-sized house, but Ashley's extended family overwhelm most houses. I sat on the ground, drinking Coca-Cola for most of the day, stealing a seat when I could.

At the end of the festivities, when my hangover subsided and we were all completely stuffed, we crowded onto the staircase. From top to bottom, children and grandchildren leaned in over the banisters to get into the picture.

It was chaos, as always. And I wouldn't have it any other way.

The Last Splash

So, here we are. If you've made it this far, well done, and thank you! We've been to Glasgow, Denmark, China and the outer reaches of Scotland. We slogged our way through a sheep rescue, survived winter in The Shire and very nearly joined the Royal Air Force.

Looking back, I didn't write this book to reflect on the things I've learned, but if you live through enough disasters, the lessons sneak up on you, whether you want them to or not. Finishing the book forces me to ask myself what I want you to take from it. And I do mean YOU. Life throws shit in your face. People say things that eat into the very fabric of your soul. It can feel like there's no way out, like you want to curl up into a ball and close your eyes for all eternity. I guess this book is my way of telling you, you are not alone.

Fuckups, mishaps and horrendous situations are inevitable. They teach us and temper us. They give us strength for the future and the confidence to overcome the next challenge, in the knowledge that each day will end, and a new one will begin. Without these trials, you'd remain fragile and weak, but with them …? With them, you grow stronger, more resilient, and adaptable in the face of adversity.

Like the kicks from those cows, these episodes are part of the game. The point isn't to avoid them. It's to find meaning in them, to laugh at them and carry on with your day, to hopefully come out of them wiser, tougher, more fully formed. You might even get a tattoo out of it.

When the shit hits the fan, you win or you learn. You've been here before and you will be again. So, collect yourself, breathe deeply and slowly. These moments of paralysing anxiety do dissipate, your heart rate will steady, and calm will return.

Life is messy. Some days it's a golden sunset with a cold beer and the love of your life by your side. Some days it's a kick in the balls from an angry heifer. Some days you won't know what it is. So, when life kicks you in the kahunas, just smile. The shit washes off, but the stories linger.

Acknowledgements

To everyone I've met along the way – whether in fields, farm-yards, train stations, or the middle of some downright sketchy situations – thank you. Somehow, we've always managed to make it out the other side either smiling, or at least a little wiser (and usually not smelling our best).

To every cow and sheep that's stood patiently, kicked unexpectedly, or reminded me that life's biggest lessons often come wrapped in muck – you've all played your part in shaping these stories.

To Gillian, our editor – thank you for your saint-like patience with my impatience, and for untangling the chaos that somehow passes for our writing process.

To Robert – this book wouldn't exist without you. Thanks for helping me piece together the timeline, confirm the details, and keep the facts straight when my memory took the scenic route.

And finally, to Ashley – thank you for still hugging me when I come home smelling of very literal code browns (even if you do make me wash at the door first). You've been my constant through every storm, scrape and shambles; my calm in the chaos, my reason for keeping going, and the only person who can somehow make even the messiest moments feel worth it.

And to you, the reader – thank you for being here, for your curiosity, and for joining me on this ongoing ride through all its unpredictable twists, turns, and occasional splashes.